C. Mattheck · H. Kubler

Wood - The Internal Optimization of Trees

Springer

Berlin
Heidelberg
New York
Barcelona
Budapest
Hong Kong
London
Milan
Paris
Santa Clara
Singapore
Tokyo

Claus Mattheck Hans Kubler

Wood - The Internal Optimization of Trees

With 121 Figures

Springer

Professor Dr. Claus Mattheck
Research Center Karlsruhe
Institute for Material Research II
P.O. Box 3640
D-76021 Karlsruhe, Germany

Professor Dr. Hans Kubler
4913 Waukesha Street
Madison, WI 53705, USA

Second printing of the first edition

The original edition (hardcover) appeared 1995 in
Springer Series in Wood Science (ISBN-13:978-3-540-62019-8)

ISBN-13:978-3-540-62019-8 Springer-Verlag Berlin Heidelberg New York

Library of Congress Cataloging-in-Publication Data

Mattheck, C. (Claus), 1947-
Wood : the internal optimization of trees / C. Mattheck, H. Kubler.
p. cm.
Includes bibliographical references (p.) and index.
ISBN-13:978-3-540-62019-8 e-ISBN-13:978-3-642-61219-0
DOI: 10.1007/978-3-642-61219-0

1. Wood. 2. Wood--Mechanical properties. 3. Trees--Mechanical properties. I. Kubler, Hans. II. Title.
QK647.M38 1997
575.4'6--dc21 96-49603

Cover design: Design & Production, Heidelberg
Cover illustration: Line drawing by C. Mattheck
Typesetting: Best-Set Typesetter Ltd., Hong Kong
SPIN 10553974 31/3137 5 4 3 2 1 0 - Printed on acid-free paper

Foreword

Here are two physicists looking over the fence of physics, getting thrilled by the life and growth of trees, taking an altogether different, exciting view of wood: trees produce wood for their own benefit. They do not live for the benefit of man who builds his world using wood as a raw material. Timber is revealed in a different light, and the reader is taught to stop thinking of it in terms of defective beams and boards. Wood only fails as a part of the living tree. To us, the tree and wood biologists, this new definition is a real, inspiring challenge, which is just what Kubler and Mattheck intended it to be. Their answers may seem too simple or little logical to some of us; but the authors are not at a loss for sound and solid arguments. Their field studies prove the incredible, their hypotheses makes us want to get to the bottom of the unproven unbelievable. The authors' answers and arguments are bold and courageous. They arouse our curiosity and force us to fathom the facts. It seems as if Kubler and Mattheck wanted to trick us into believing that trees only live and react following mechanical rules and strategies. To tell the truth, that was what I first suspected the authors of; but I was wrong. Kubler and Mattheck have never denied the role of the climate and of the soil, of water and light, of nutrients and air pollutants, of hormones, and other messengers which control the growth of trees. They have never doubted their importance, but have simply left that field of research to the tree physiologists. The vivid style used by Kubler and Mattheck avoids complex formulas preferring impressive drawings. Kubler and Mattheck, a successful pair of authors, whose wisdom, wealth of ideas, and spontaneity have given them the courage and heart to open new perspectives, should go on peeping through the gaps in the fence. They should keep on inspiring us and attracting our attention by doing so. They have reopened blocked-up approaches to a better understanding of the trees which surround us. The tree and wood biologists should not leave them alone in their efforts, but should take over part of the work. If joint action is taken, we will surely be rewarded, and will soon profit from more information about trees as the teachers of nature.

Prof. Dr. Dieter Eckstein
Institute for Wood Biology
University of Hamburg

Foreword

There is something fascinating about picture books: you leaf through them, your eye is caught by an attractive illustration, and you are suddenly spellbound and cannot put the volume aside before you have seen the rest. That was just the way I felt when I started to review the manuscript submitted to me by Kubler and Mattheck. I do not know much about physics; to tell the truth, I dread formulas; but the bird shown in the first two figures boosted my self-confidence by chirping a soft but clear and encouraging "you-understand-the-mechanical-fundamentals-I'm-sure-you-do" in my ear.

Kubler and Mattheck intend to introduce a general readership to the perfect physics of the trees in our surroundings. The language the authors use is simple, and the pictures they present are illustrative and a pleasure to look at. The physical background could not be understood so well by readers who are unfamiliar with physics if Kubler and Mattheck had insisted on complex elucidations instead of giving basic, intelligible explanations and examples of the observable characteristics of trees:

- The lateral branches of trees are as short as possible or they bend into the direction of the wind to prevent wind blows from destroying the trunk. The tree combines its perfect physical and biological features in the constant effort of maintaining an even load distribution.
- A glance at material that the tree attaches to close wounds or to integrate foreign bodies shows that extreme stress concentrations are missing during the tree's lifetime.
- The authors convinced me that tree rings and wood rays have a physical function that completes the acknowledged biological properties. The tree rings and wood rays are arranged so perfectly that critical destructive shear and lateral stresses are controlled.
- Many a wood-defining feature, for example, enlarged wood rays in the area of the tree ring boundaries of beeches, suddenly turns into and is recognized as an appropriate, a vital physical feature. The refreshing, impressive drawings made me look at wood anatomy in an altogether new, unconventional way.

Mattheck is not only good at drawing, and Kubler and Mattheck do not intend to present a populistic masterpiece. Both authors are experienced scientists who have a profound knowledge of physics, and who convince by their technical skills and enormous didactic qualities. I could wish that teachers of this kind had educated and guided me through my studies.

Much remains to be investigated, many fibers remain to be drawn. For all the findings introduced, tree biology still keeps some secrets, and Kubler and Mattheck are well aware of that. Biologists, chemists, and physicists may unite to get to the bottom of the unsolved mysteries during the course of time. This book paves the way for such future action.

Reverence for nature and modesty about our achievements obtrude upon our minds, knowing that nature invented the first tree in the Devonian period; 365 million years later we are trying to understand the fundamentals, not the details, of this ingenious, successful invention. Why? Certainly not just for fear of heavy, cracked branches falling down on us as we go for a stroll through the town on a windy day in autumn. The reader himself may find the answer.

Prof. Dr. F.-H. SCHWEINGRUBER
Swiss Federal Institute for Forest
Snow and Landscape, Birmensdorf, Switzerland

Preface

To use a piece of wood in a truly optimal way, we have to expose it to the same load conditions the tree faced when that particular part of the tree was growing and being invented. Whoever studies isolated dead timber samples looks at an optimized structure, but does not know for which purposes it had been optimized.

For example, when you tension-load a piece of timber which grew in the tree under compression load, this is misuse of material. Or if you straighten a bent piece of wood, which during the tree's life was even more bent, this is misuse of material. It is close to sabotage when you ignore the tree's knowledge.

To learn what the tree knows, it is very helpful, if not an absolute must, to use equipment that can measure mechanical wood properties under field conditions, face to face with the tree. For that, *Fractometers* were invented. This procedure allows easy access to the strength and stiffness distribution within a tree. The sample required by them is an increment core, easily removed from the tree with an increment borer, by contrast with the alternative process of felling the tree, sawing out timber, and finally machining samples for each feature under study. This alternative and traditional research method is time-consuming. Even worse, it may lead to alteration of the material properties; for example, drying stresses may develop before measurements can be completed.

Our focus will be on the biomechanical function of the woody structure, not on anatomical details. Hopefully, the wood anatomist will excuse this approach, as well as some simplifications in the use of nomenclature. Loss of details is the price to be paid for simplification and functional clarity. The book offers a number of new insights on the internal optimization of wood that now await their biological explanation.

The authors are grateful to the people who contributed to the book, especially to the coworkers in the biomechanics department at Kernforschungszentrum Karlsruhe. Mrs. H. Knierim made the English translation and Mrs. A. Kubler-White read the proofs. Mrs. D. Gräbe exercised angelic patience in preparing manuscript details. Prof. D. Eckstein (University of Hamburg) kindly checked the manuscript from the viewpoint of a wood biologist.

The management of Kernforschungszentrum Karlsruhe had confidence in the success of our work and followed it with warm interest and steady support. Last, but not least, the authors thank all the foresters and friends worldwide who supported the field research.

The authors would also like to express their thanks to the Editor of this Series, Prof. Tore Timell, for many improvements to the text and his generous support of this book and the helpful team of Springer-Verlag for dealing with all the numerous details until final production.

Karlsruhe, Germany
Madison, USA
August 1995

CLAUS MATTHECK
HANS KUBLER

Contents

1 Introduction

There is a wide choice of good books about the biology of trees, and much excellent work has been written about timber and timber-derived products, and about the anatomy of wood, for example Carlquist (1988), Schweingruber (1990), and Wagenführ (1989). Most of the work on mechanical properties of wood is done in a vacuum, ignoring its use within the tree itself. The tree manufactures its wood in a very load-adapted way in order to prevent mechanical failure.

The present book is meant to be a topical, systematic guide to the rules and laws of tree design. The reader is introduced to the design rules that control the inner architecture of trees. New findings are discussed and proven. The optimum use of wood by the tree and in the tree is described in terms of the tree's ways and means of varying the biomechanical properties of its wood to its best advantage. The failure mechanisms that would destroy the tree if such adaptation and optimization were neglected are discussed. Optimization prevents failure!

All the newly established theorems are proven by presenting the results of theoretical calculations and experimental field studies. The knowledge founded on facts and figures is supported and completed by some unproven but plausible speculations and assumptions. Special measurement methods and instruments were developed to determine the quality of the wood of green trees. These new methods and instruments are described. The latest significant development among instruments, the FRACTOMETER, is introduced. Many of the results presented could never have been obtained without it.

This book intends to show the adaptation of the tree's outer shape and inner architecture to unavoidable external stresses as induced by, e.g., gravitation or wind loads. Since the underlying processes cannot be understood without a certain minimum of mechanical knowledge, the essential mechanical fundamentals are presented in the first section, which is a slightly modified version of a passage found in Mattheck (1993). Experienced engineers may skip it. Biologists and foresters will find it a helpful preparation for all further sections.

2 Mechanical Fundamentals

2.1 External Loading and Internal Stresses

Mechanical components are exposed to various external loads such as forces, bending moments, torsional moments, and thermal stresses. To avoid undesired movement, the stationary components of the tree must be provided with bearings or supports that compensate these loads through reaction loads.

This is easily understood if one regards the example of axial forces (Fig. 1). (While readers with some good basic technical knowledge may skip this passage, readers with a biological background may find it diverting to be introduced to the necessary technical terms through examples which are observed in nature.)

Assuming the mass of a rather well-fed bird to be m, one obtains a weight $F = m \cdot g$, where g is the acceleration of gravity. To simplify matters, the weight of the pile on which the bird chose to settle is neglected. The weight of the bird effects an elastic compression of the pile in the axial direction. This elastic compression is characterized by a uniform compressive-stress distribution $\sigma = F/A$ (A: cross-sectional area of the pile). The soil which bears the pile must produce an equal but opposite force F to avoid that the pile sinks in. Axial tensile forces cause tensile stresses in the pile and produce the corresponding contrary effect.

Matters are rather more complicated in the case of eccentric loading (Fig. 2), which occurs when the well-fed bird settles on a lateral branch. The crown of the tree is assumed to have broken, and the trunk and the branch are imagined to have no weight. Emphasis is on the load of the bird, on its mass. The weight of the bird, F, acts on a lever arm, l, at some distance from the trunk. The bending moment increases linearly from the bird to the branch joint, where it reaches its maximum $M_B = F \cdot l$. An equal but opposite bending moment M_B must be produced by the tree to support the branch. Since other forces are missing and the lever arm remains unchanged due to the verticality of the trunk, a uniform, constant bending moment acts downstem. This bending moment must be compensated for by the roots in the soil; otherwise, in the worst imaginable case, the tree turns over and the bird is wary of trees forever. The compressive stresses which the weight of the bird $F = m \cdot g$ (Fig. 2) induces in the case of eccentric loading act in the trunk alone.

One and the same bending moment can induce bending stresses of different intensities. Thin lateral branches are exposed to higher stresses than thick trunks. This is explained in the following. The bending stresses are defined in

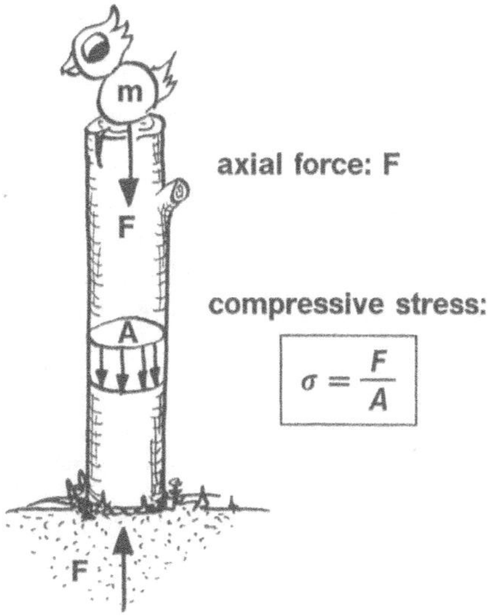

Fig. 1. The weight of a bird whose mass is m introduces an axial force F = m g (g: acceleration of gravity) into a pile. The compressive stresses $\sigma = F/A$ (A = cross-sectional area) that act in the pile are caused by this axial force

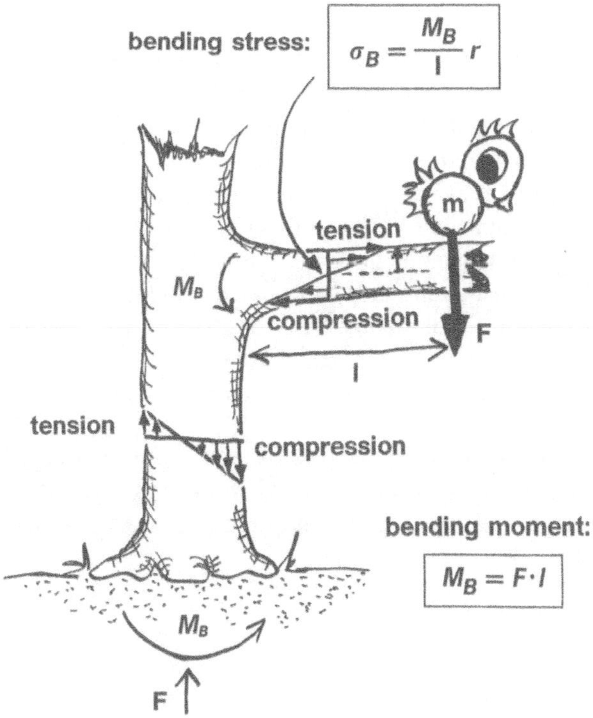

Fig. 2. When the bird settles on a lateral branch, the branch-to-stem bending moment increases linearly in the branch and remains constant as it acts downstem, adding up to the weight load. The branch is subjected to pure bending stresses, while the stem is exposed to bending and pressure because it carries the weight of the bird in addition (see Fig. 1)

accordance with Eq. (1) if circular cross sections are assumed for both the branch and the stem.

$$\sigma_B = \frac{M}{I} \cdot r. \tag{1}$$

Evidently, the bending stresses increase linearly from the center to the surface of the stem. They act as tensile stresses in the positive and as compressive stresses in the negative r-direction. Equation (2) applies to the circular cross section.

$$I = \frac{\pi}{4} R^4, \tag{2}$$

where R is the radius of the beam and I is the geometrical moment of inertia, which characterizes the shape of the cross section.

If Eq. (2) is entered in Eq. (1), the maximum bending stresses which act on the beam surface are

$$\sigma_B^{max} = \frac{4M}{\pi R^3}, \tag{3}$$

which means that in spite of equal bending moments, a stem that is twice as thick as a branch is exposed to only one eighth of the stresses as the branch.

Equation (1) and the stress distributions shown in Fig. 2 reveal a problem which arises from the specific distribution of the bending stresses. In the center of a bending bar ($r = 0$) the stresses are zero, i.e., there are neither tensile nor compressive stresses. The load capacity of trees in this underloaded area may be excellent; however, the wood which one finds there is wasted material now as regards the tree's ability to resist bending. While nature evidently tolerates such waste in the case of anchored, unmoving trees, it provides the moving mammals with bones which are mostly hollow in the area of the neutral fiber of bending ($r = 0$), thus avoiding material which serves no load-carrying function.

The intelligent use of material by trees in spite of their naturally limited possibilities becomes evident in the case of unilateral wind loads. The stem ovalizes in the wind direction and develops an elliptic cross section whose geometrical moment of inertia is

$$I = \frac{\pi}{4} a \cdot b^3, \tag{4}$$

with (a, b) being the semiaxes of the ellipse.

The tree attaches all its building materials where the bending stresses are highest by developing thicker tree rings on the tensile and compressive sides. While any such attachment is raised to the third power of the greater semiaxis b [Eq. (4)], wider tree rings in the direction of the small semiaxis would be relevant only in the linear sense. The oval cross section which the tree develops proves to offer the greatest resistance to the preferential bending load. Stresses in the oval cross section are lower than those in a circular cross section that is equal in area and is exposed to the very same external bending moment. Some examples of root cross sections throughout this book remind one of I-beams, which are known from civil engineering (Fig. 3). Such roots are formed when the tree ceases to add more wood in the area of the neutral fiber of bending (r = 0).

Just as trees adapt their shapes, they must subordinate themselves to the rules of nature which give priority to bending loads and force biological components to shape-optimize in many curious and complicated ways, without survival guarantees.

While axial tensile and compressive loading can be treated separately in theory from bending, in practice they rarely occur in pure forms. Figure 4 illustrates combined tension and bending. The weights of the trunk and the branch are neglected again to simplify the example. Uninfluenced by the wind, loading is assumed to be only by the weight (due to the force of gravity) of the upper and lateral crowns (F_1, F_2). Only the weight of the upper crown produces compressive stresses $\sigma_1 = F_1/A_1$ by acting in the stem above the lateral branch. The lateral branch is exposed to a bending moment which increases linearly towards the trunk (see Fig. 2). Therefore, the trunk below the branch joint must bear both the bending moment that is induced by the lateral branch

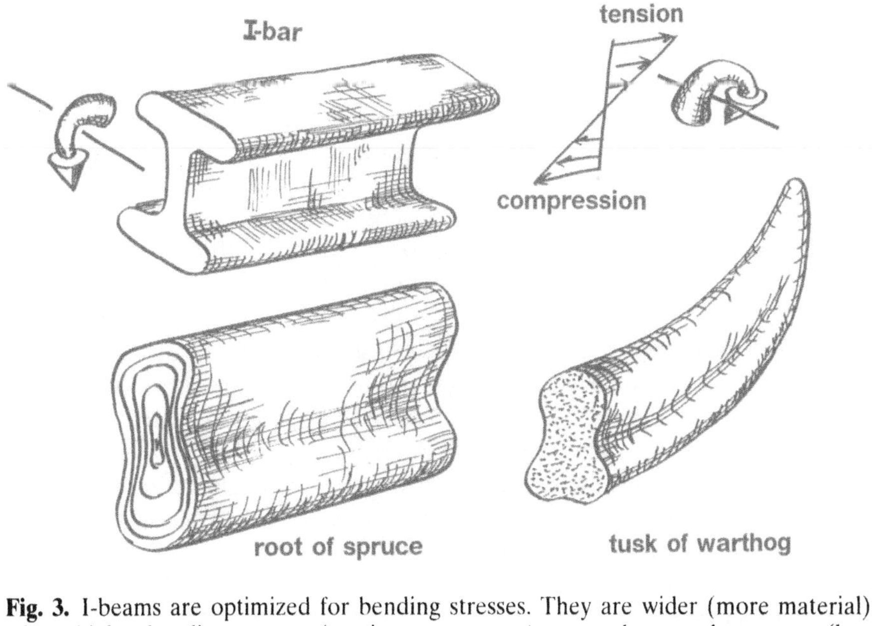

Fig. 3. I-beams are optimized for bending stresses. They are wider (more material) where higher bending stresses (tension or pressure) act on them, and narrower (less material) where there are lower stresses. Root cross sections and the tusks of the warthog are examples of optimized I-beam shapes found in nature

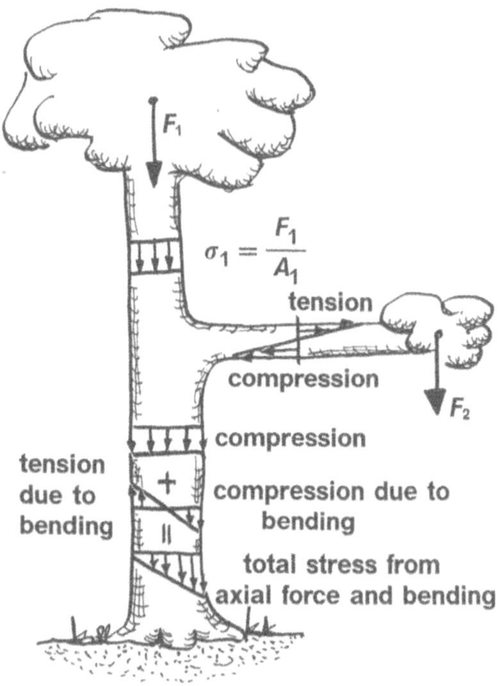

Fig. 4. Combined axial stresses and bending stresses induced by the crown weights F_1 and F_2

(linear distribution of bending stresses), and an axial compressive stress that is increased by the vertical force F_2 originating in the lateral branch, but is transmitted to the larger lower circumference of the stem.

Elastic stress analysis leads to the combined stress distribution by simply adding the partial component stresses. The left side of the stem exhibits the tensile stresses of bending which are quasi-neutralized by the axial compressive stresses due to $(F_1 + F_2)$. On the right lower side of the stem the axial compressive stresses and compressive bending stresses simply add up. If a wind-bending load were added, one would superimpose the resulting distribution of wind-bending stresses. This is discussed in greater detail in the sections below.

The tensile and compressive stresses, and the bending stresses (mostly linear transition from tensile to compressive stresses) may be accompanied by shear stresses (Fig. 5), which act tangentially in shear-loaded areas and prevent the separated parts from yielding.

Shear stresses also act in the middle layer of bending bars, i.e., on the neutral fiber of bending. The pages of a book that are bent slip relative to one another in a similar way. Qualitatively, the shear stresses differ widely from the tensile stresses and bending stresses which act perpendicularly to the reference surface.

As is evident from Fig. 6, the shear stresses produced by torsion in the cross-sectional plane of a twisted cylinder increase linearly from the center

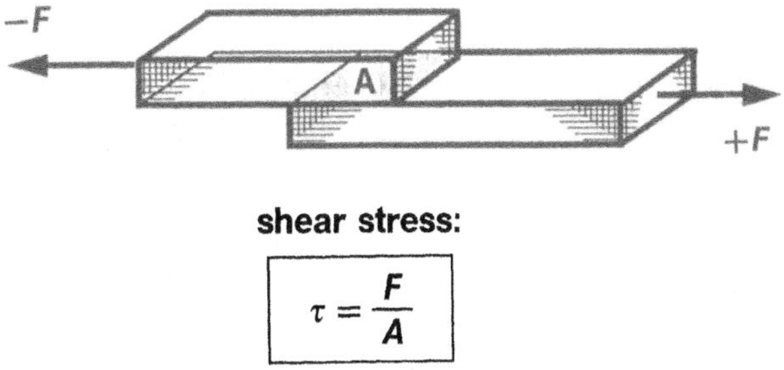

Fig. 5. Shear stresses act tangentially to the reference surface. They prevent the separated parts from yielding

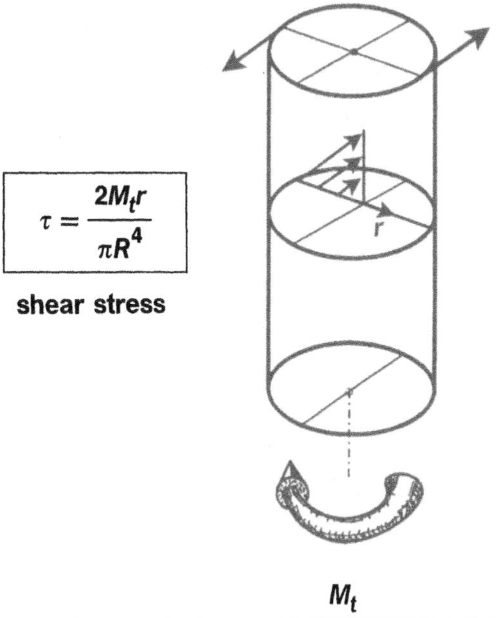

Fig. 6. The shear stresses acting in the circular section of a twisted cylinder increase from the center to the envelope

(r = 0) to the envelope (r = R) of the cylinder. These shear stresses, which are induced by a torsional moment M_T, can be superimposed with other types of elastic stresses. Applying these facts to trees, stems with unilateral crowns and with branches extending towards the light may be twisted under wind loads and be exposed to shear stresses in addition to the loads described above.

Things are not as easy as they may seem because, normally, loads are multiaxial, i.e., there is more than one stress component per load case. Different stresses act in different directions; for example, braces contract transversely as they are pulled longitudinally. Since it would be beyond the scope of this book to discuss all the components of the stress tensor, interested readers should refer to Currey (1984) for further details.

Although a simplified number of stresses would make an easier study, their multiaxial character cannot be denied completely because of its relevance to growth simulation, discussed below. For this reason, a closer look is taken at the principal stresses σ_1, σ_2, σ_3 (Fig. 7). An imaginary small cube is

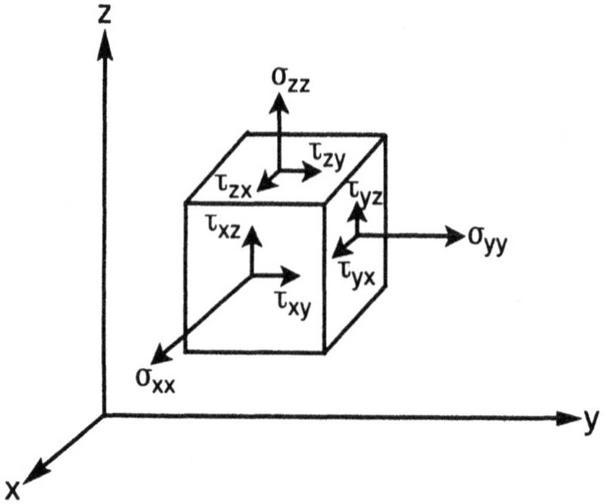

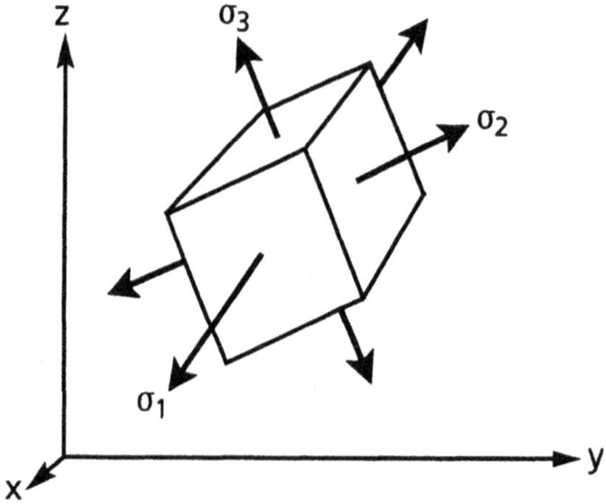

Fig. 7. When the cube has a certain orientation, normal stresses act on its surface while shear stresses (shown *above*) are absent. These normal stresses act along the principal stress trajectories (*below*)

inserted in the component so that stresses σ act perpendicularly to the cube surfaces while tangential shear stresses τ are completely absent. The "normal", perpendicular stresses represent the principal stress trajectories σ_1, σ_2, σ_3. Their vectors are often referred to as the "force flow". If the cube were pulled in one direction ($\sigma_1 > 0$) it would contract in the remaining two directions through compressive transversal-contraction stresses (σ_2, $\sigma_3 < 0$).

A reference stress is defined to assess the overall mechanical loading of the cube without any need to regard the individual values of the principal stress trajectories. The Mises reference stress

$$\sigma_{\text{Mises}} = \frac{1}{\sqrt{2}}\sqrt{(\sigma_1 - \sigma_2)^2 + (\sigma_2 - \sigma_3)^2 + (\sigma_3 - \sigma_1)^2} \qquad (5)$$

is a failure stress which is widely used to predict plastic flow in steel. It is introduced because it is relevant to the simulation and control of biological growth.

Simply speaking, the Mises reference stress is a measure of the cube's exposure to pulling and pushing from all sides. Growth simulation (Mattheck 1993) profits by the application of the Mises stress on account of the square root [Eq. (5)]. The Mises reference stress is always positive, i.e., programming can be done ignoring preceding signs, even if positive tensile stresses may change into negative compressive stresses. The deformation of components under external loading must not be neglected during this excursion into internal stresses.

Figure 8 illustrates the correlation between external loading and component deformations in the case of uniaxial tension. The strain $\varepsilon = \Delta l/l_0$ which results as one applies a force F for deflection Δl of a tension bar whose length is l_0 is defined as a relative change in length. A linear correlation is assumed to exist between the stress $\sigma = F/A$ and strain $\varepsilon = \Delta l/l_0$ in the case of small deformations and elastic material behavior. For this one-dimensional case one obtains Hooke's law

$$\sigma = E \cdot \varepsilon, \tag{6}$$

where E is the factor of proportionality which is referred to as the modulus of elasticity or Young's modulus.

The Young's modulus of steel is $E_S = 200\,000\,\mathrm{Nmm}^{-2}$, while that of green wood is $E_H = 10\,000\,\mathrm{Nmm}^{-2}$ along the grain. Anisotropic biological materials such as wood or bone are characterized by directional variations in their Young's moduli.

Nevertheless, just to simplify matters, we assume biological materials to be isotropic. Critical readers may frown on this idea but should reconcile themselves to it, considering the following: biological structures have their main force flow running along the grain, an adaptation of their inner architec-

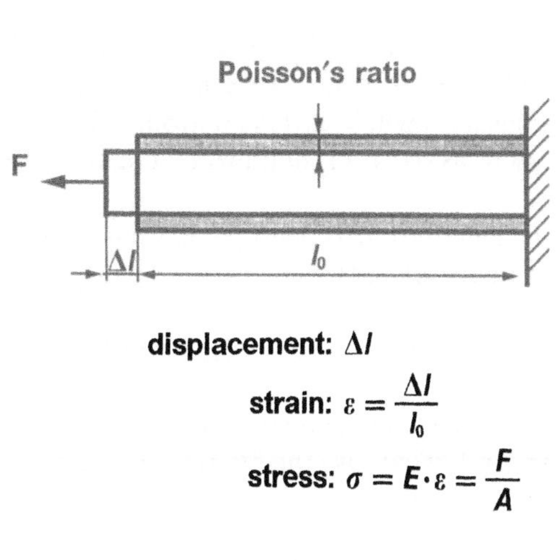

Fig. 8. Hooke's law elucidated by the axial forces, stresses, expansions, and displacements assumed for a tensile bar with cross section A

ture. Since the Young's moduli along the grain vary only insignificantly, it is acceptable to equalize them, so to speak, and cheat for the sake of elucidation as regards the not-so-relevant lateral stresses. The good agreement between theory and nature justifies this trick, and shows that it makes sense in this case to assume homogeneous, isotropic materials for the purpose of a clearer presentation. However, where nonisotropic behavior cannot be avoided, it will be regarded in this book especially when optimum fiber arrangement is calculated.

2.2 Thermal Expansion and Thermal Stresses

Figure 8 also serves to explain the thermal stresses in components. Neglecting any lateral expansion, the external force, F, is replaced by the temperature difference ΔT which heats the bar and causes a change Δl in its length. The bar remains unloaded if allowed to extend freely. Thermal stresses would build up, however, if elongation were impeded by clamps or walls at both ends of the bar. Thermal expansion

$$\varepsilon = \frac{\Delta l}{l_0} = \alpha \Delta T, \tag{7}$$

(α = coefficient of thermal expansion) and Hooke's law [Eq. (6)] combine into

$$\sigma = E \alpha \Delta T, \tag{8}$$

which defines the thermal stresses.

The coefficient of thermal expansion is a material-dependent constant, to be found experimentally for each component one wishes to investigate.

Thermal stresses are not only caused by mechanical restrictions but may also occur without these, on account of internal temperature gradients. Biological growth can be simulated on the basis of similar processes of thermal expansion (Mattheck 1990, 1993). This so-called CAO method (Computer Aided Optimization) simulates biological growth by stress-controlled thermal expansion. It is already widely used as well in optimizing engineering structures.

The mere knowledge of the static equilibrium, of Hooke's law, and of the correlation between expansion and displacements can provide appropriate solutions to simple mechanical problems.

However, the ingeniously elegant and functional designs which make biological structures survive in nature are too complex to be described applying the usual mechanicotheoretical methods. Any analytical approach that is based on formulas is bound to fail miserably because the interrelationships of nature are not caught by formulas. For example, the mechanical conditions on the surface of an elaborate bone cannot be pressed into a feasible mathematical framework. Before today's sophisticated numerical computer methods came on the market, the shapes of components were approximated by applying simple geometries (e.g., circles, straight lines, parabolic curves, ellipses). The resulting long, terrifying equations were unreliable, and their further reduction into quick, rough estimations caused even more confusion. The finite-element method, which is discussed below, is one of several altogether novel approaches to component analysis.

2.3 The Finite-Element Method (FEM)

The component analysis methods and mainframes which have been on the market for a few decades are truly revolutionary achievements. The finite-element method (FEM) is an incredibly successful analytical tool which has developed during this time. FEM is an efficient aid in component analysis; almost all the results introduced in this book were obtained through applying it. FEM has been winning popularity among design engineers, and has been used for stress analyses of tires, propellers, skis, even saucers, and other articles of daily use. Engineers can tell a tale or two about frustrations in finite-element practice. However, much as they may curse the difficulties with which they grapple, they have confidence in FEM, and find it irreplaceable. Many a design problem could not have been solved without this method, and growth simulation would never have been invented without FEM.

The finite-element method is based on a relatively simple principle. Equilibrium conditions are neglected, while the displacements, extensions, and stresses that characterize mechanical component loading are inferred from minimum-energy data.

The selected component is decomposed into finite geometrical segments, the "finite elements" (Fig. 9). These elements are defined by their edge points and coordinates. The actual shape of the component is represented by the whole of the elements. There are no limits to the complexity of this shape. One of the great advantages of FEM over other methods is its ability to provide appropriate solutions regardless of the form and shape of the component under examination.

After making the FEM model, one defines the material properties of the individual elements, i.e., their Young's moduli, coefficients of thermal expansion, and Poisson's numbers, which describe the transverse contraction. While conventional analytical methods would capitulate to variations among material data of the element, FEM tolerates differing values and permits solutions in spite of them. In a next step, basic service conditions of the components are laid down, and their bearings, clampings, supports, or guides are determined (for example, the four legs of a chair must rest on the floor, vertical displacements are not allowed). A specific load is defined, differentiating between line loads, surface loads, and concentrated loads (Fig. 10).

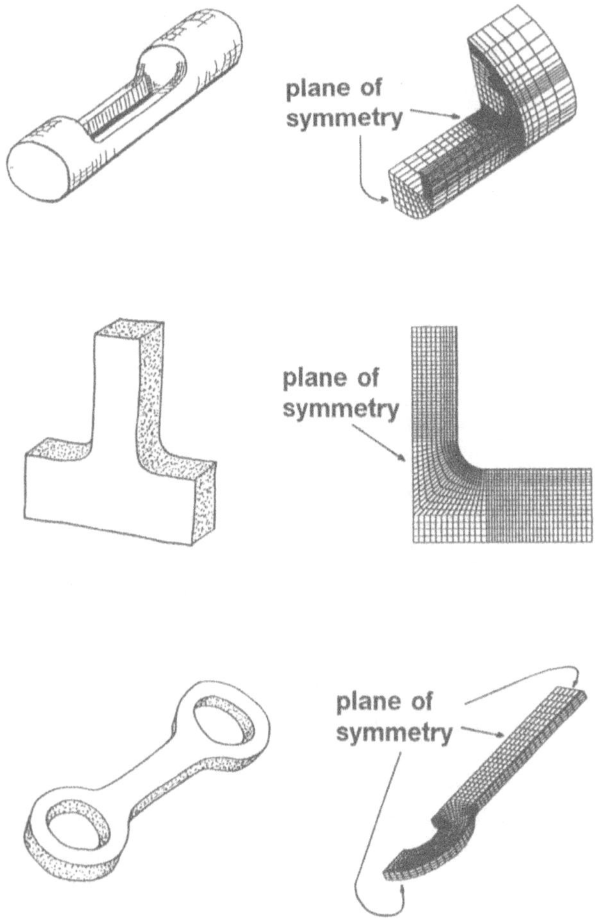

Fig. 9. Some mechanical components and their idealized finite-element models. One half or one fourth of the entire structure provides a sufficient basis for stress analysis of a symmetrical component with one plane or two planes of symmetry. Rotationally symmetrical structures can be represented as plain longitudinal sections whose rotational symmetry is defined by means of additional inputs

Several or single parts of the contour can be displaced to investigate reaction stresses of the component.

The input data compiled in the course of these preparations are entered in an FEM computer program for calculation of the component's most efficient energetic state. This procedure, which is just another way of determining the static state of equilibrium, gives very exact information about the displacements, strains, and stresses at each point of the component. This mechanical approach is completed by a physical modeling step that reduces and digests the data and facilitates the interpretation of results. Most of today's commercial

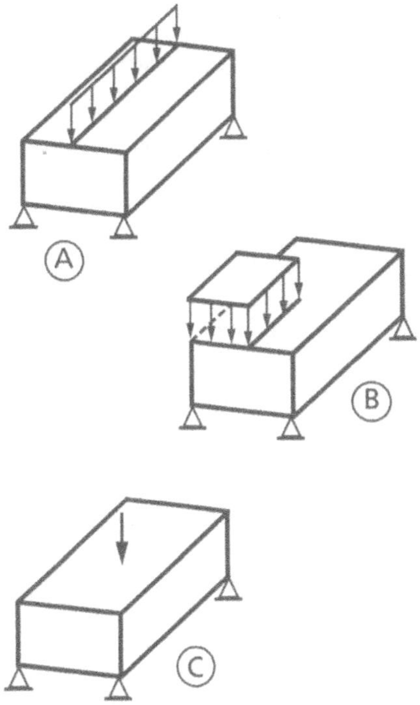

Fig. 10A–C. Basic mechanical load and service conditions with line loads (**A**), surface loads (**B**), concentrated loads (**C**), and specified edge displacements

FEM programs provide high-precision solutions and are suited for personal computer implementation. Further information is given by Mattheck (1993).

The stress analysis of a notched component, which is discussed below, will familiarize prospective users with the application and practice of FEM. Notches and notch stresses are component killers whose significance should not be underestimated.

2.4 Notches and Notch Stresses – The Component Killers

Figure 11 shows the stress distribution in a tensile plate with a circular center hole. The stress σ_0 that is applied acts at some distance above and below the hole. The inserted hole has a double effect. It reduces the width of the plate by its mere presence, and causes higher stresses, $\sigma_1 > \sigma_0$, even if one assumes uniform stresses over the weakened cross section.

The effects of a normal nonoptimized notch are worse and less predictable than those of the presented model notch.

The force flow that approaches the notch from the top and bottom must be diverted around the notch, as with a flowing liquid. The force flow surrounding the notch normally effects higher stresses. These are the so-called notch

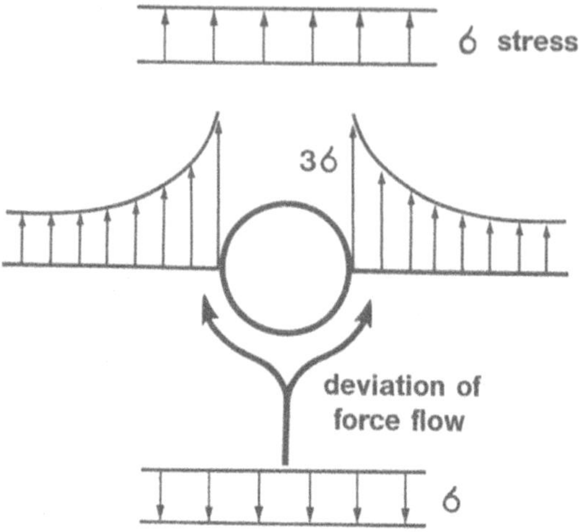

Fig. 11. Failures are often caused by culminant local notch stresses. Stresses near the circular hole are at least three times as high as anywhere else in the tensile plate. The stress intensity depends on the distance between the hole and the edge of the plate

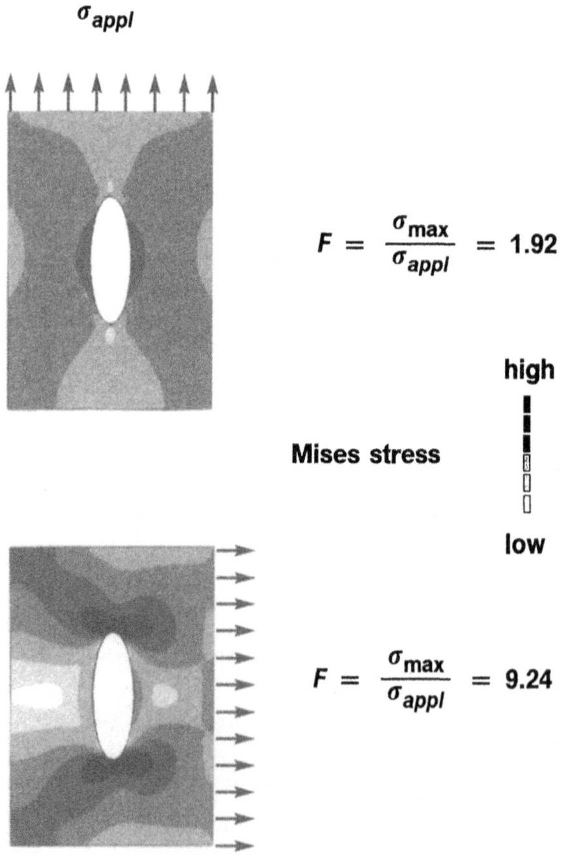

A

Fig. 12A–C. Different notches that exhibit different notch stresses but result in equal reductions of the cross section of the plate. F is the increase in the stresses applied externally. **A** Elliptical notch. **B** Parallel slot with semi-circles at end. **C** Circular notch

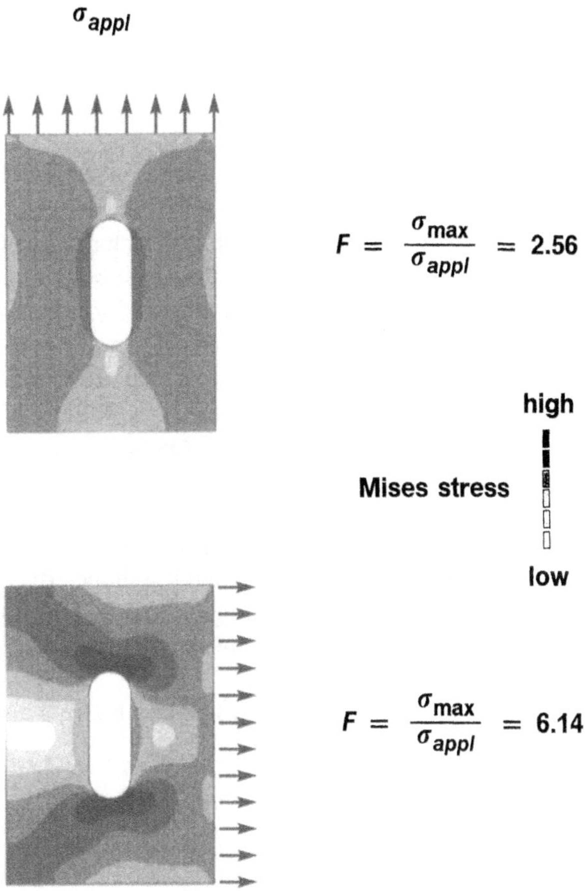

σ_{appl}
$F = \frac{\sigma_{max}}{\sigma_{appl}} = 2.56$
high
Mises stress
low
$F = \frac{\sigma_{max}}{\sigma_{appl}} = 6.14$

B

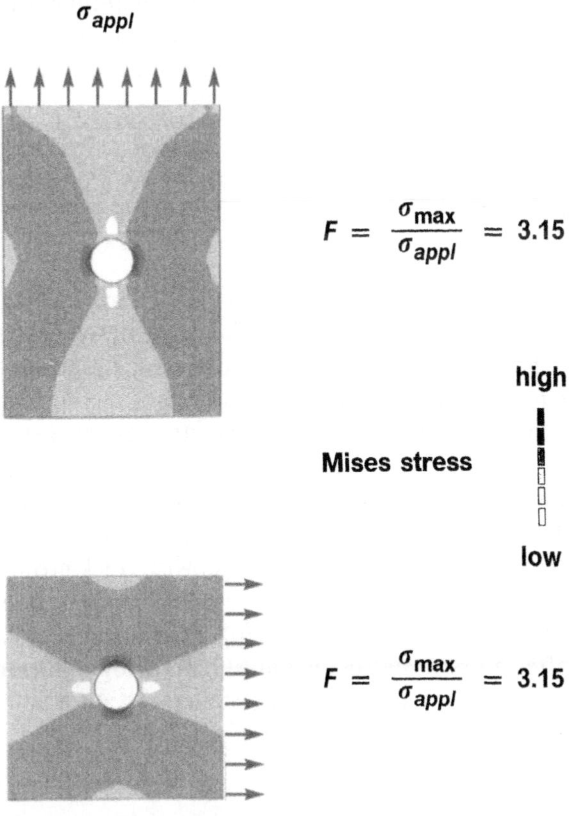

σ_{appl}
$F = \frac{\sigma_{max}}{\sigma_{appl}} = 3.15$
high
Mises stress
low
$F = \frac{\sigma_{max}}{\sigma_{appl}} = 3.15$

C

Fig. 12B,C

stresses. The intensity and severity of the notch stresses depend on the roughness or abruptness of force flow diversion. Figure 12 shows that the effect of a notch is often essentially determined by the orientation of the notch relative to the load direction (Mattheck 1993). Although the force flow in solid components establishes a static distribution of load, it may be compared to the flow of liquids. Just as with liquids, a notch whose shape and surrounding flow resemble a ship's bow will be "circumnavigated" by a force flow more smoothly, i.e., with lower notch stresses, than a notch of identical shape but exposed transversely to its load. Any notch reduces the cross section by the notch width. Irrespective of the shape of a notch, the stresses σ_1 which are determined by the residual cross section are identical. The local increase in stresses at the edge of a notch, however, is determined exclusively by the shape and by the abruptness of the specific force flow diversion that the shape effects. While circular notches remain neutral, a pointed longitudinal notch cuts through the force flow more smoothly and elegantly than the pointed ends of a transverse notch (Fig. 12).

F is a factor which represents the relation between maximum stress and applied stress:

$$F = \frac{\sigma_{max}}{\sigma_{appl}} . \qquad (9)$$

This form factor, which is sometimes defined as $\sigma_{max}/\sigma_{appl.}$, determines the increase in stresses due to the presence of a notch in the component. It is a measure of the critical nature of a notch with regard to component failure (Peterson 1974).

2.5 Crack Propagation

Figure 13 illustrates the failure of a component by example of a nonsymmetrical circular notch in a tensile plate. The force flow presses through the bottleneck on the left between the hole and the edge. The stress produced there is much higher than the hole edge notch stresses on the right-hand side. If one exposes the component to cyclic axial stresses for a sufficiently long period of time, and carefully enough to avoid overload breakage, the left edge breaks first and the crack slowly propagates. The final separating sudden rupture occurs as the crack reaches the stage of unstable crack growth. One speaks of fatigue crack growth in the case of slow, stepwise crack growth, and refers to unstable crack propagation in the case of the final sudden fracture. Overload breakage is the term which is used when unstable crack propagation occurs right after a first excessive loading. The radiograph (Fig. 14) illustrates the serious, critical consequences of fatigue cracks initiated by notches, seen in some implants fractured in human bodies.

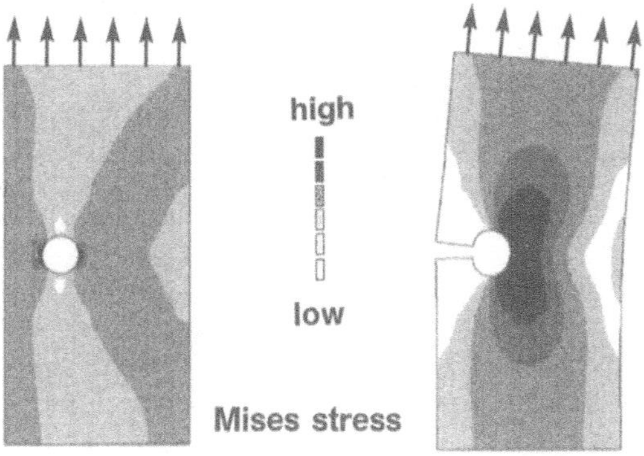

Fig. 13. Very high near-edge notch stresses cause components to fail through the action of cracks that start on one side of the hole (*circular notch*) and propagate after rupture of the narrow edge strip. The computer plots reveal the enormous increase in Mises stresses in the area of the crack tips and at the edge of the circular notch

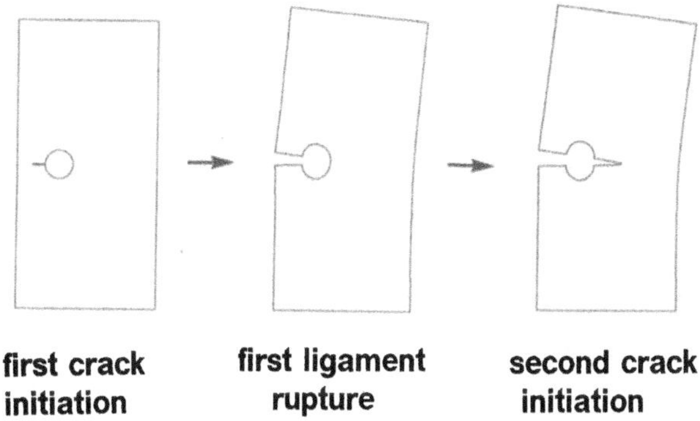

The systematic survey given below emphasizes and summarizes the main mechanical aspects in a less narrative way.

2.6 The Main Mechanical Aspects

- Elastostatic-problem formulation is based on the following input data:
  Component geometry
  External loading (forces, surface loads, line loads, bending moments, torsional moments, internal pressure; Fig. 15)
  Boundary conditions (bearings, clampings, guides)
  Material data (Young's modulus, Poisson's ratio v, coefficient of thermal expansion α)
- These input data are entered in a computer program, e.g., FEM, for calculation of the component stresses, strains, and displacements.
- The external loading is transmitted into the bearings. The bearings compensate the stresses acting on them through equal but opposite reaction forces.

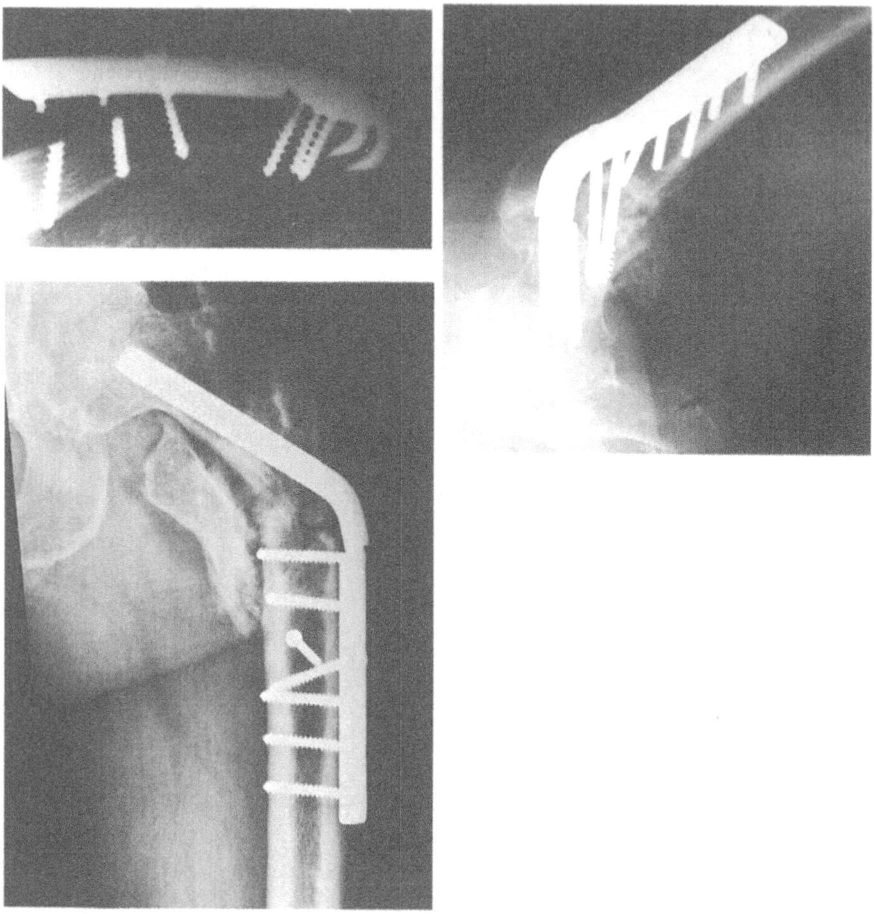

Fig. 14. Implant fatigue fracture initiated by notch stresses

- Thermal stresses develop when thermal expansion is impeded by external clampings. They also build up in the presence of temperature gradients, which effect conflicting differences in thermal expansion at different points in the component.
- The Mises reference stress,

$$\sigma_{\text{Mises}} = \frac{1}{\sqrt{2}} \sqrt{(\sigma_1 - \sigma_2)^2 + (\sigma_2 - \sigma_3)^2 + (\sigma_3 - \sigma_1)^2}, \tag{10}$$

  is a measure of the multiaxial loading at each point in the component. The Mises stress in components characterized by a preferential load direction quasi-coincides with the greatest principal stress value.
- Notch stresses develop through force flow diversion in the concave areas of the outer component contours (e.g., screw threads). They also occur in the

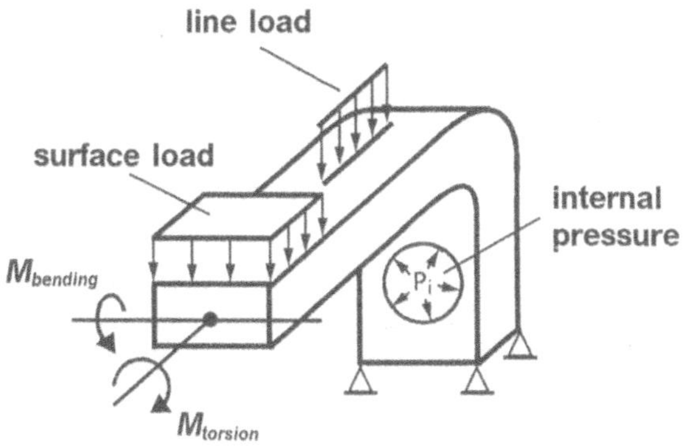

Fig. 15. Multiple external loading, and component bearings

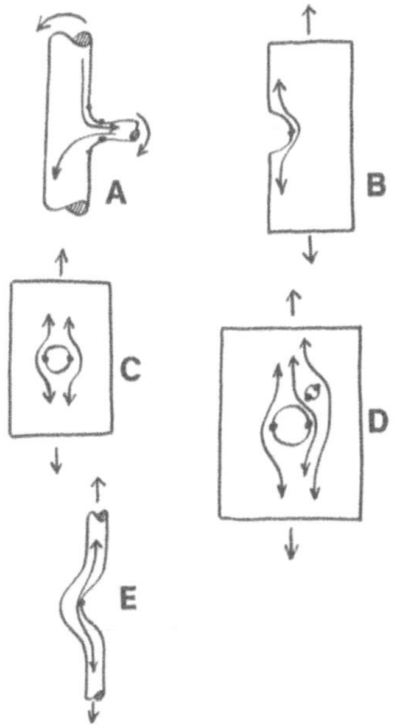

Fig. 16A–E. The concavity of notches always effects a diversion of the force flow (Mattheck 1991). The *dark dots* represent the points of highest notch stresses

presence of internal notches such as bore holes. The severity of a notch is defined by the increase in stresses expressed through $F = \sigma_{max}/\sigma_0$, where σ_{max} is the maximum notch stress and σ_0 a reference stress far away from the notch in an unaffected part of the component. Figure 16 shows a variety of notches, all of which are concave relative to the contour of the component.

- Notches may be completely harmless in one load direction, yet highly destructive in another load direction, even under insignificant loading. This decisive influence of the shapes of notches is essential to component optimization through "component growth", with the aim of determining notch shapes that do not induce notch stresses. (Notches in living structures normally do not cause notch stresses; Mattheck 1993.)
- Under cyclic loads, fatigue cracks may be initiated where the notch stresses

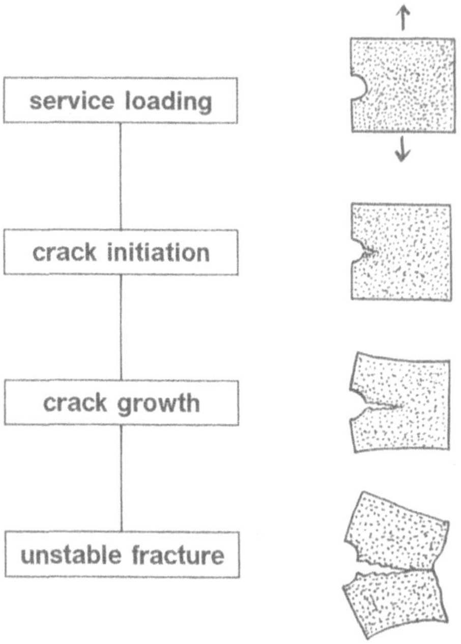

Fig. 17. A chain of cause and effect: from component loading to fatigue fracture

are highest. Each time the load is applied anew, stable fatigue crack growth continues. It becomes unstable when the crack has reached a critical length. This final unstable fatigue crack growth inevitably destroys the component. Cracked defective components must not necessarily be rejected, provided one fixes reliable inspection intervals which avoid the fatal consequences of unstable crack growth in accordance with the principles of fracture mechanics.

The chain of cause and effect shown in Fig. 17 illustrates crack growth and final fatigue fracture. While the previous sections have described stress development in components, notch stress formation through force flow diversion, fatigue crack initiation, and fatigue fracture, the following chapter discusses approaches to controlling and avoiding premature component failure.

3 The Mechanical Self-Optimization of Trees – Five Theorems

The mechanical fundamentals discussed above lead to five substantial theorems. Unproven yet convincing, these theorems will be the reader's guiding hypotheses through a labyrinth of amazing phenomena and intricate details. A summary at the end of this book invites the reader to judge the truth and value of the arguments presented and to form his own opinion.

3.1 The Five Theorems

- The principle of minimum lever arms
  The tree minimizes external stresses by reducing the length of the loaded lever arm. Minimum lengths can be achieved through active self-bending of the stiff parts (reaction wood!) or through passive yielding of the flexible parts of the tree.
- The axiom of uniform stress
  Unavoidable stresses are distributed evenly on the surface of the tree. Stresses on the surface of the tree are uniform at each point on the time average; grown notches do not cause notch stresses.
- To minimize critical shear stresses, the wood fibers and tree rings are arranged along the axial or tangential force flow.
- The strength of wood depends on the distribution of the mechanical stresses that act in the tree when it is exposed to external loads. The outer shape of the tree *and* the local internal quality of the wood are optimized and adapted to the degree and type of loading.
- Unavoidable weak points are counterbalanced by the growth stresses developed by the tree. The growth stresses counteract the critical loads that may cause failure. The distribution and degree of growth stresses are stress-controlled.

The first two theorems, which determine the outer shape of the tree, are discussed only briefly. The reader may refer to Mattheck (1991, 1993) for further details. The main part of this book is dedicated to the remaining theorems, i.e., to the conditions and processes that control and determine the inner architecture of the tree. Examples are given to show that the violation of any of these five mechanical "rules of conduct" can ruin the tree or lead to the failure of vital parts. The observance of the hypothetical theorems established above enables the tree to survive.

4 The Principle of Minimum Lever Arms

4.1 Negative Gravitropism

The traveler who waits for his train at the station is not expected to carry his heavy suitcase with his arm outstretched, and the women of primitive tribes walk along carrying their heavy buckets on their heads. Both the tired traveler and the experienced women reduce the lever arm, which is the point of application of the force of gravity.

The trees resort to a similar method (Fig. 18). After the loss of the leading shoot, the outstretched, lateral succeeding branch straightens to take the place of its predecessor. The potential successor's urge to straighten is suppressed by the ruling leading shoot as long as it is intact. This suppressing, distance-demanding command from the leading shoot to side shoots is referred to as apical dominance. Apical dominance and negative gravitropism are balanced continuously to adjust the optimum angle between branch and stem. These two corrective mechanisms may be assisted by phototropism. When the leading shoot ceases to assert its claim to apical dominance, phototropism and negative gravitropism take action, and control the shape of the tree by encouraging the straightening efforts of the most qualified side branch. This branch, mostly, but not necessarily, the uppermost branch, then asserts its claim to apical dominance and suppresses the remaining potential leading shoots. Equally matched candidates may jockey and fight for the leading-shoot position over years or even decades.

Trees naturally leaning, so to speak, on slopes straighten themselves through formation of reaction wood. This inner mechanism of reaction wood formation will be discussed in detail later. Figure 19 shows a leaning tree which, depending on the taper of its stem, grows into a saber tree or develops a meandering shape. In both cases (Figs. 18, 19), negative gravitropism succeeds in reducing the bending stresses, especially those that act on the bases of the trunks.

4.2 The Flexibility Strategy

In addition to the active negative gravitropism that shortens the loaded lever arms intentionally, so to speak, there are also passive regulating mechanisms. Grass and flowers are bent and flattened by the wind, and man instinctively

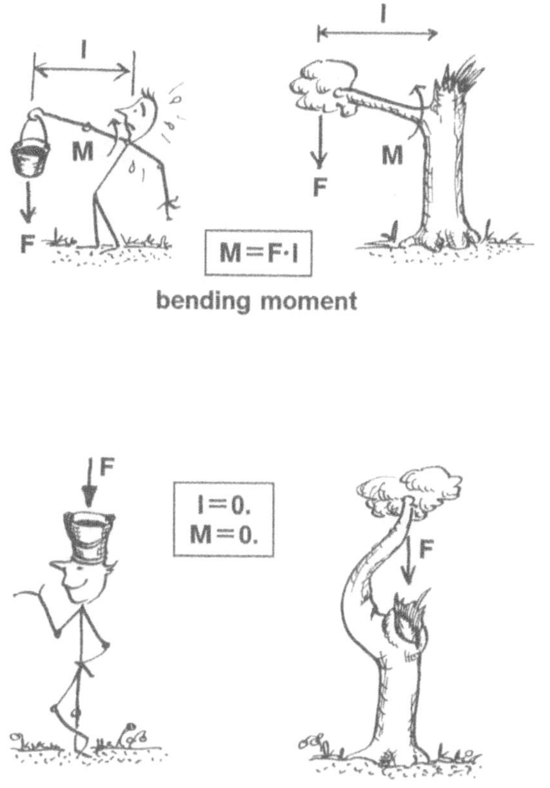

Fig. 18. A lateral branch straightens after the loss of the leading shoot, thus minimizing the bending load that acts on the branch joint and in the stem

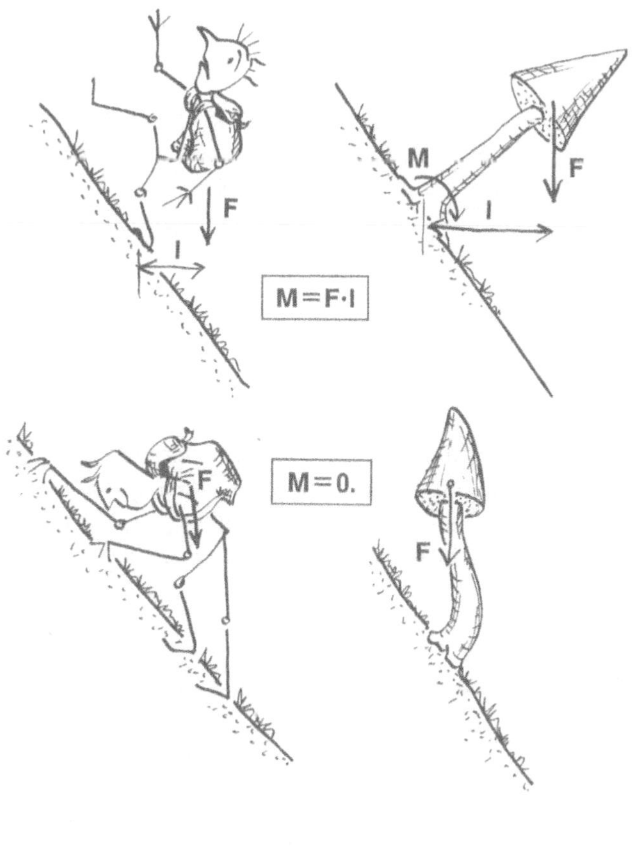

Fig. 19. Saber trees or meandering saber trees minimize the butt-bending moment

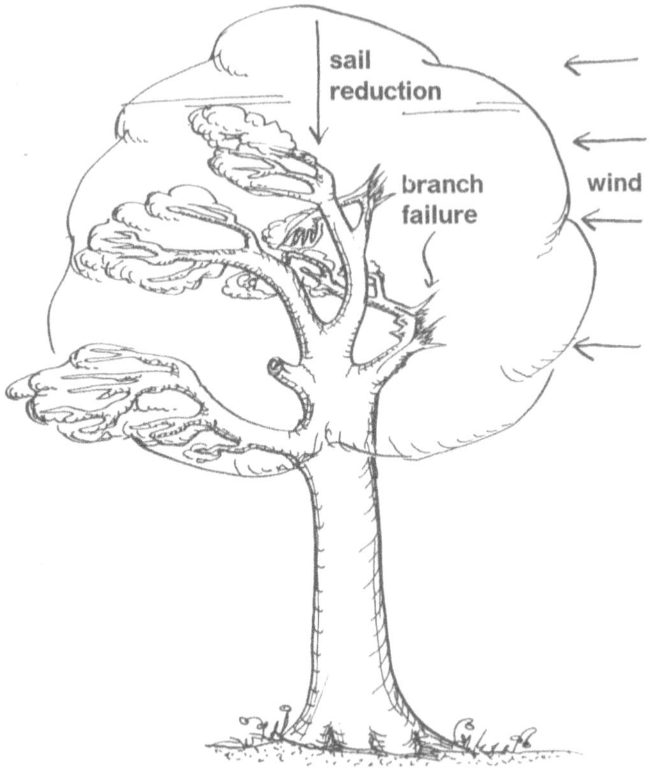

Fig. 20. The flexibility strategy applies to the peripheral parts of the tree

stoops to reduce the "area of the sail". Trees, or rather the fringes of the crowns of trees, are found to apply a similar flexibility strategy (Mosbrugger 1990). They reduce the area of their sail by bending twigs and even leaves in the wind direction. The loss of branches in stormy weather may be considered a "safety valve", a protective branch-before-stem breakage mechanism (Fig. 20). The tree survives so long as the stem does not break, hence branches may be lost for the sake of the preservation of the species. Sailors caught in heavy storms on the open sea take in the reef and do not cut the mast except to survive, i.e., unless they recognize that losing the mast is the only way to keep the hull functional.

Technical valves are designed to prevent vessels from bursting; they open to avoid pressure buildups. The safety valve function of tree branches is fulfilled only when their safety factors are smaller than the safety factor of the stem that carries them. The safety factors of stems are slightly above S = 4.5 (Mattheck et al. 1993) if one assumes the ratio

$$S = \frac{\text{failure load}}{\text{service load}}.$$

While the failure load is defined by the design and material strength, the maximum service load depends on the external loading to which the tree is

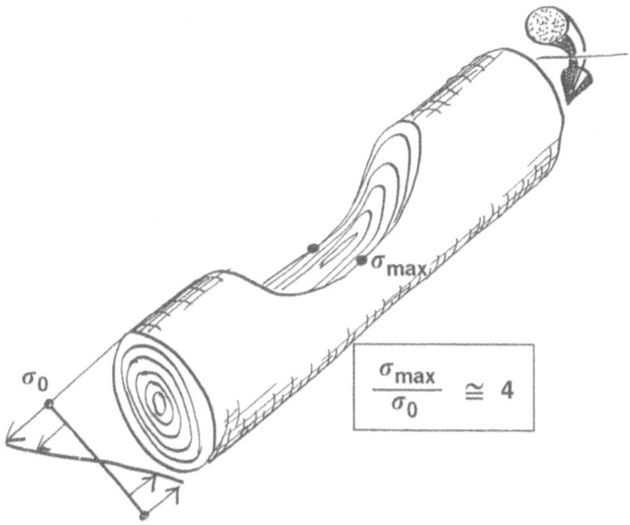

Fig. 21. Experimental notching for determination of the safety factors of branches

exposed. Field studies gave safety factors of around S = 4.0 for branches. In one study of branch breakage, notches of different sizes and shapes were cut in a large number of branches of different species and of different diameters. The notch shown in Fig. 21 caused some of the branches to break, yet many of them remained and healed the wound. The safety factor of stems $S \geqslant 4.5$ was determined by applying a similar field study method. The rather small difference between the safety factor of stems $S \geqslant 4.5$ and that of branches S = 4.0 seems most appropriate considering that greater discrepancies would allow any gentle breeze to deprive the trunk of its vital branches. The difference between $S \geqslant 4.5$ and S = 4.0 is not too great, yet not so small as to prevent the tree from sacrificing its branches in time to save the trunk.

The safety factors of mammals are in the range of S = 3–4 (Alexander 1981, Currey 1984). Nature has chosen these factors in view of the fact that animals move about carrying the whole weight of their biomass. Each excess pound of bone makes them less competitive, consumes energy, and increases the chance that they will lose during flights or fights. The hiker on a long walk prefers his light and easy-to-carry rucksack to heavy bags, and the traveler thinks twice about taking three suitcases if there is no taxi to carry them to the train station. Since trees, however, do not move about, they can tolerate a few pounds beyond their optimal minimum.

The minimization of lever arms achieved by the tree by bending its twigs and branches in the wind direction is assisted by yet another reaction that completes its repertoire of flexibility strategy. This mechanism is controlled by the difference between the safety factors in stems and branches, respectively. When bending has reached its limits, the tree resorts to lever arm reduction by breaking branches. Although this is not a truly flexible, elastic reaction, it is a flexibility strategy in the wider sense. The tree takes in the reef by rejecting branches, and sets sail again by growing new sprouts. This flexible reaction to

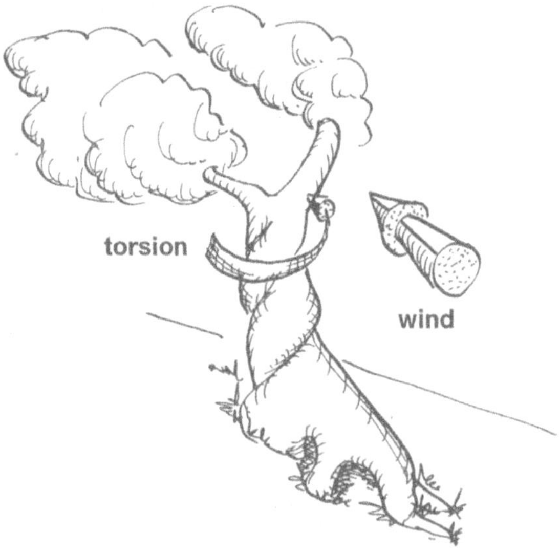

Fig. 22. Spiral grain as the result of one-sided torsional loading over a period of many years

wind loads by the tree crown needs considerably more time to grow new branches than for the instantaneous elastic movement of the branches.

Mechanically controlled spiral grain through asymmetrical crowns (Fig. 22) is another manifestation of the flexibility strategy. Spiral grain may reduce lever arms permanently. When a color bearer faces the wind holding his flag with arm outstretched to one side, he is turned about on his axis. Applying this effect to trees in the open, one finds that the wood fibers are arranged along the spiral force flow induced by torsion. This phenomenon will be substantiated and discussed in detail later. When permanent spiral grain is formed, the protruding lateral branch is bent behind the stem. This elastic, momentary minimization of the lever arm develops into the permanent wind load protection of the tree over the course of time. Wind beeches in Scotland are often found to exhibit this spiral-grain phenomenon. To further reduce heavy wind loads in stormy weather, their branches may finally break in spite of the twisted stems. Unlike broken trunks, broken branches can be replaced!

5. The Axiom of Uniform Stress

5.1 The Undisturbed Design

A German forester (Metzger 1893) found that the tapering typical of spruce trees will ensure a uniform distribution of the surface bending stresses, if the stem diameter, D, is cubical relative to the distance, h, from the effective point of application of the wind load (Fig. 23).

The sudden change in stem diameters in the area of branch joints is a vivid example of the load-distributing, into-the-ground downstem fiber arrangement and force-flow orientation of the branches. Above and below the branch joints one finds stresses of equal intensity.

The axiom of uniform stress manifests itself conspicuously in the area of the roots. According to Mohr-Coulomb's law, the shear strength of the ground increases in proportion to the degree of shear surface compression. This increase may be compared to the increase in friction proportionately to the growing pressure of the friction surfaces against each other. The wind load that acts on the tree is caught by the crown (sail) and passes the trunk (mast) to enter the ground. The ground is compressed on the lee side and lifted on the windward side (Fig. 24). Regardless of the fact that the shear strength on the lee side increases while that on the windward side is reduced, the weaker side of the ground must bear almost equally intense stresses because the weight acting on the ground is negligible compared to the wind force that acts higher up. The less shear-resistant ground on the windward side must be reinforced by additional, longer roots, which are not necessarily found on the stronger lee side. Since the roots on the windward side are exposed to higher loads, the bending stiffness of their cross sections is for the most part greater than that of the roots on the lee side.

The axiom of uniform stress can also explain the shapes of root cross sections as being the sums of all tree ring width distributions. Figure 25 shows an I-beam-shaped root as the result of pure bending loads. It grew strongest on the highly stressed compression and tension sides of bending and developed least in the middle, i.e., in the area of the neutral fiber of bending, where bending stresses are absent. Buttress roots preferably develop on the windward side (Mattheck 1991) in the presence of sinker roots that join the horizontal roots at a greater distance from the stem. They are the result of superimposed tensile stresses and bending stresses that add up on the upper side and are partly or completely neutralized on the lower sides of the roots. It is due to the resulting decreasing stress distribution that the upper tree rings

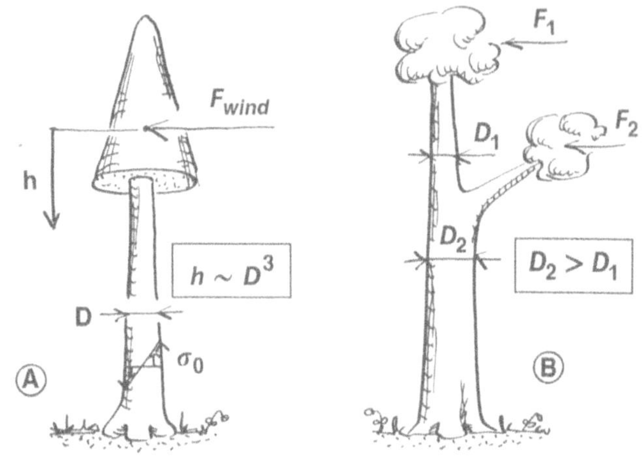

$$\sigma_0 = \frac{F_{wind}\, h}{\frac{\Pi}{32} D^3} \stackrel{!}{=} constant$$

$$\longrightarrow \quad h = \frac{\Pi\, \sigma_0}{32\, F_{wind}} D^3$$

Fig. 23A,B. Stem-tapering defined by the requirement of uniform bending stresses. **A** Tapering of spruce stems (Metzger 1893). **B** Local change in diameter due to lateral introduction of the wind load by a main branch into the stem

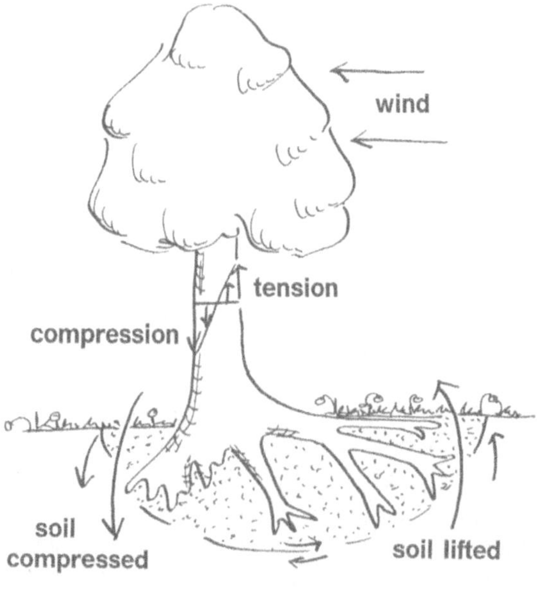

Fig. 24. Root morphology defined by the uniform stress axiom and by Mohr-Coulomb's law

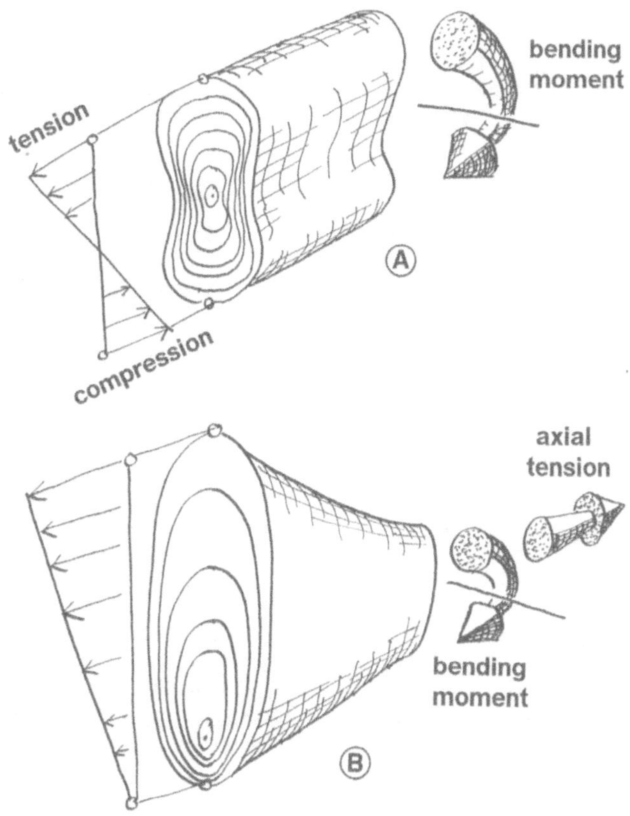

Fig. 25A,B. Root cross sections tell the load history. **A** Pure bending leads to I-bar formation. **B** Maximum stresses on the upper and zero stresses on the lower side of the root lead to buttress root formation

are wider than the lower ones. Tree ring thicknesses, thus, are records of the stress distribution inside the ultimate tree ring. While the average circumferential incremental growth depends on the prevailing growth conditions, the circumferential distribution of the tree ring thickness is controlled purely mechanically. Tree rings are records written in the body language of trees. They can neither be distorted nor falsified, and are valuable, nonappealable, and incontestable documents when produced in evidence during lawsuits.

Avoidable loads are avoided through negative gravitropism or through flexibility strategies, while unavoidable loads are distributed evenly in accordance with the axiom of uniform stress. When the biomechanical optimum achieved in this way is disturbed by external influences, the tree resorts to self-repair in order to restore the optimum and regain its ideal state.

5.2 Repair Growth

A circular hole punched in a tree causes a force-flow diversion that is often accompanied by locally high stresses (notch stresses). Only naturally grown

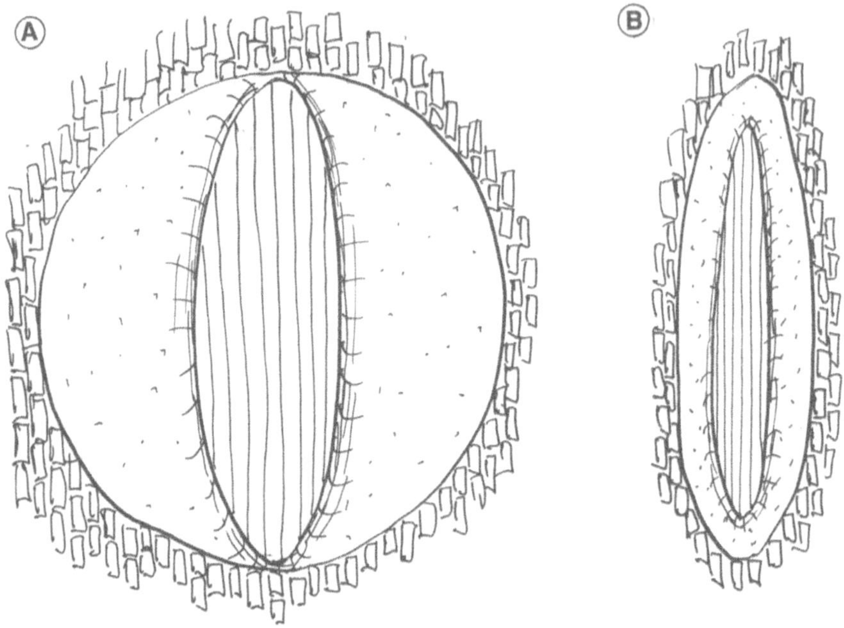

Fig. 26A,B. Wound occlusion is fastest where notch stresses are highest. **A** A circular hole heals into a spindle-shaped notch whose longitudinal axis points in the direction of the force flow. **B** A spindle-shaped notch punched in the tree right at the beginning heals equally fast at any point

notches are free from notch stresses. The circular hole shown in Fig. 26 exhibits maximum notch stresses on both sides. These stresses accelerate the progress of wound healing by letting the circular hole heal into an oval spindle whose greater semiaxis points in the direction of the force flow.

When a spindle instead of a circular hole is punched in the tree from the beginning, one finds that stress concentrations are missing and that wound healing is equally fast at any point of the spindle. Arboriculturists who wonder how to cut the edges of a wound left after the collision of a car with a tree at the roadside are taught by the tree to cut the rugged edges into an enveloping spindle. No matter how complex or strange a wound may be, the tree consistently repairs it where stresses are highest. The wound-wood of two opposite or adjacent edges may meet and weld into a common tree ring. Figure 27, drawn from photographs, shows some examples of such formations.

Callus collars, formed by trees to surround dead or insufficiently vital branches (Fig. 28), once more reveal that material is attached where it is actually needed. These collars are annular notches that serve as predetermined breaking points; they help the tree to survive by inducing useless parts to break. According to finite-element calculations (Mattheck 1993), the edge of the collar coincides with the high-to-low-load interface that separates the stem from the dead branch. Evidently, the collar is the boundary contour of

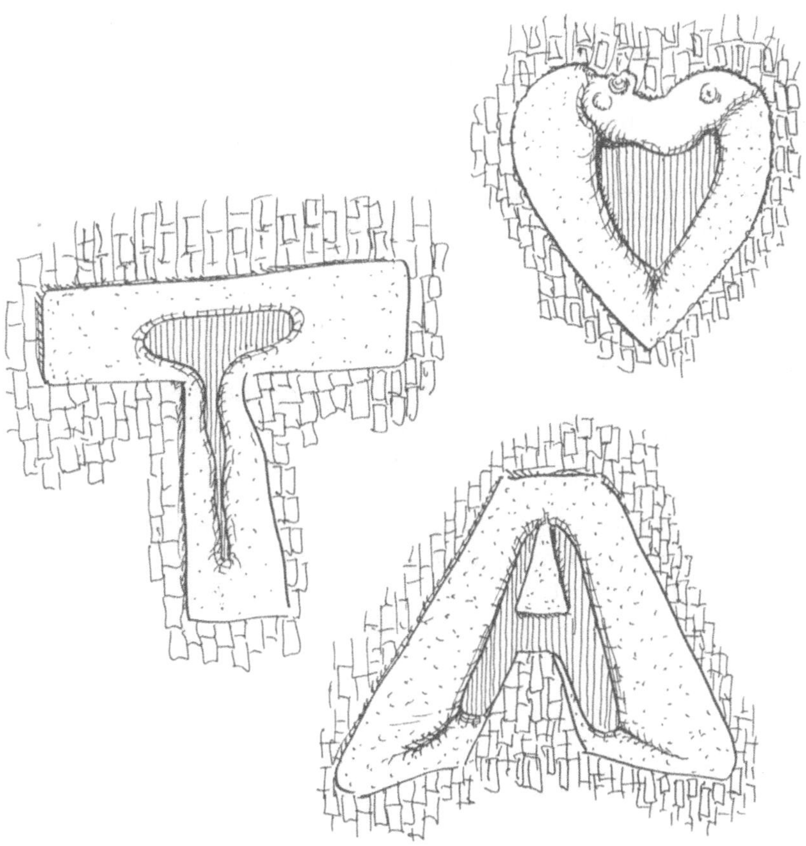

Fig. 27. Letter- or heart-shaped wounds heal fastest where notch stresses are highest, i.e., where the axiom of uniform stress is fulfilled least

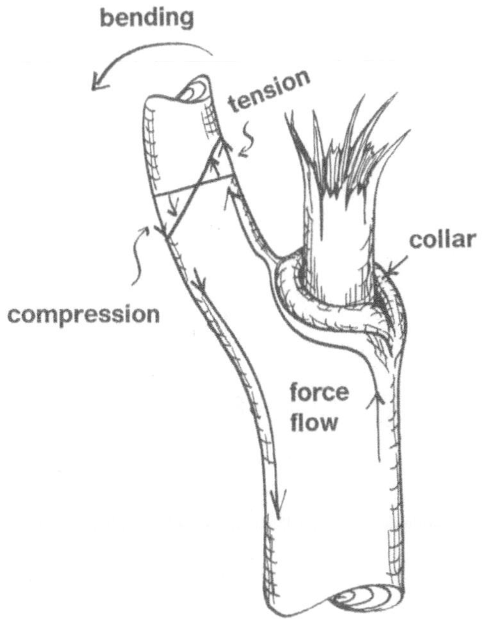

Fig. 28. The collar with which the living parts of the tree surround the neighboring dead parts separates the loaded from the nonloaded area

the loaded growing part, i.e., the wood attachment area of the tree. The top view (Fig. 28) shows that the collar is shaped like a spindle.

The state of even stress distribution can also be disturbed by mechanical contacts, i.e., by surface pressure. This phenomenon is depicted and dramatized in the fairy tale, The Princess and the Pea, where the princess feels the contact stress of a small pea despite heaps of cushions and eiderdowns, thus proving her delicacy, her sensibility, her true identity, and noble birth to the prospective prince consort.

Cushioning for the purpose of contact stress reduction is also observed in the case of trees. A root (Fig. 29) anchored under the weight of a stone enlarges its contact surface by developing a knotty shape that improves and reinforces its hold in the ground. It seems that roots prefer such "social contacts". They entwine round stones, they twist around neighbors, they weld and melt to avoid wear or open wounds that would increase the risk of decay.

Cushion formation is the more pronounced the greater the concentration of stresses. Figure 29 shows the vivid example of a tree growing around a steel

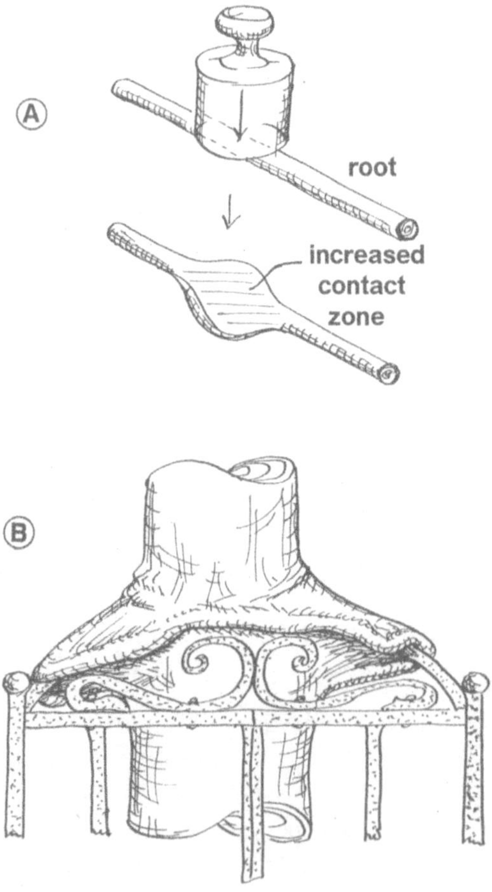

Fig. 29A,B. Contact stresses are reduced by enlargement of the contact surface. **A** Local root enlargement under the pressure of a weight. **B** A poplar grows over a steel railing. The contact area is enlarged due to high contact stresses along the line load

railing. The formation of cushions and enlargement of contact surfaces serve to reduce notch stresses and avoid permanent wear. The sawn cross sections shown in Fig. 30 reveal that the tree rings are arranged perpendicular to the contact surface. This perpendicular arrangement minimizes the shear between the tree rings. The partial stems of tapering tree fork or separate, but close-set trunks tend to press against each other on account of their secondary thickness growth. They are contacting foreign bodies, each of which enlarges its contact surface. Two stems of the same species can thus weld, provided there are no kinks or bends to impede the smooth and gradual fusion of their contacting tree rings. Cross-weldings are nothing unusual, but they will not be discussed here. Theoretical descriptions and further examples may be found in the literature (Mattheck 1991, 1993).

In addition to the above qualitative description of shape optimization in accordance with the axiom of uniform stress, two examples are given below to elucidate computer-simulated repair growth. CAO (Computer-Aided Optimization) was developed for the simulation of stress-controlled biological growth.

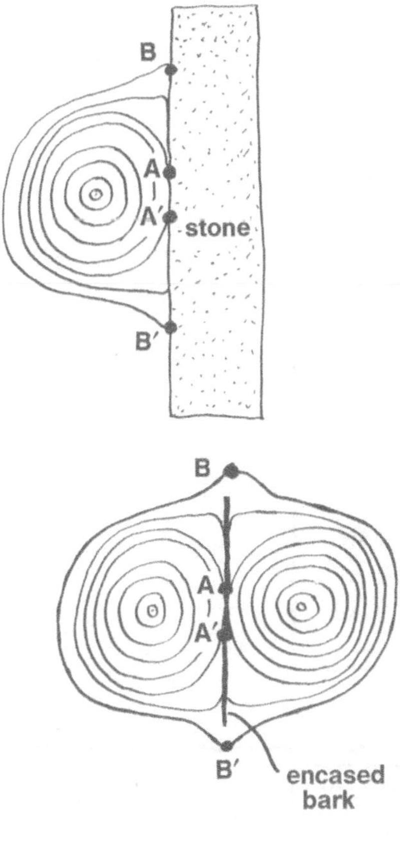

Fig. 30. Under axial pressure, the contact surface increases from *A-A'* to *B-B'*. Two contacting trees both increase their contact surfaces to form a common, welded tree ring

5.3 Computer-Aided Optimization – CAO

Trees "detect" local stress peaks, and enlarge their current tree rings exactly where the higher stresses are located. The numerical simulation of this simple mechanism of biological body care, attaching materials in higher-loaded areas, allows the engineer to predict growth processes theoretically. Figure 31 illustrates the selected simulation procedure (Mattheck 1993).

This computer-aided method for optimizing the shapes of mechanical and biological components owes its remarkable success to imitation of nature: adopting the properties of the cambium where growth occurs in overloaded areas, while stagnation is found in nonloaded areas.

The CAO method was developed at the Kernforschungszentrum Karlsruhe. It has been tested on more than 300 biological and technical components and has been adopted by the German industry and by many institutes and companies abroad.

Figure 31 elucidates the steps of the procedure of the CAO method.

1. A finite-element structure of a first design proposal is generated, considering the entire range of the functional aspects that are relevant to the operation of the component in service. The FEM structure should have a surface layer of equally thick elements, at least in the area intended for "growth". This layer of finite elements corresponds to the cambium.

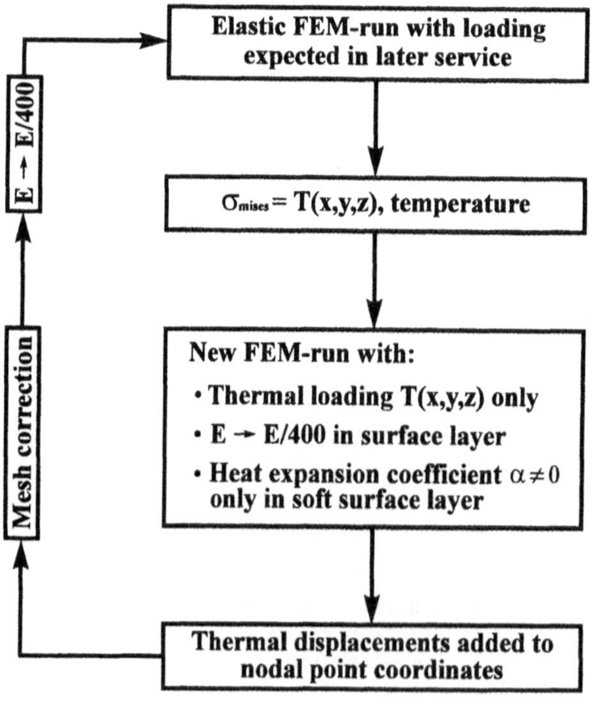

Fig. 31. CAO flow sheet

2. The intended future service loads and bearing conditions are entered, and the first FEM run is started. This first calculation gives the individual nodal displacements, the strains and stresses, and the Mises reference stress provided that one enters Eq. (5). The notch stresses in the component are thus revealed, and another run is started to prepare the simulation of adaptive growth.
3. The calculated Mises stresses are formally equated with a fictitious temperature distribution. In applying this trick, one crudely ignores the different units of measurement of stress and temperature. The trick is helpful, nevertheless, because it converts the areas of highest mechanical stresses into areas of highest temperatures in the component. In addition, one reduces the Young's modulus in the upper layer to about 1/400 of the initial value to obtain a soft surface with high-temperature overloaded and low-temperature nonloaded areas.
4. The next run finally simulates adaptive growth. The thermal stresses are considered while the mechanical stresses (tension) are set to zero. The soft upper "cambium" layer is given a coefficient of thermal expansion of $\alpha > 0$, while the solid material below is not allowed to expand. During this thermal-stress run, the soft upper layer expands or "grows" relative to its temperature distribution. The areas of highest stresses (step 2) turn into the areas of highest temperatures, i.e., into areas of maximum expansion and most pronounced growth. Since the upper layer is "as soft as butter", collisions with adjoining elements are avoided. Without being in one another's way, all elements grow inside out almost perpendicularly to the surface. This stress-controlled thermal growth

   $$\sigma_{Mises} = (T - T_{ref})$$

   can be multiplied by an adequate enhancement factor in case the calculated growth is not intense enough for the intended purpose. Extremely distorted FEM networks can be corrected. $T_{ref} = \sigma_{ref}$ is the component's desired overall service stress. The software needed for such corrections is available through the Kernforschungszentrum, Karlsruhe.

   It is recommended that potential CAO users apply a method also developed at Kernforschungszentrum, Karlsruhe: in a third run, prescribed displacements are entered by superimposing the calculated thermal growth on the structure in the absence of both thermal and mechanical stresses. The entire network is shifted gradually but does not lose its homogeneity. The results obtained by applying this method are more accurate than those that are influençed by major element deformations.
5. The four steps described above may suffice for the generation of an improved, shape-optimized structure. To ascertain whether another cycle is required, the Young's modulus of the soft upper layer is readapted to that of the base material, and steps 2 to 4 are repeated, applying mechanical stresses only. This iteration distributes the stresses homogeneously and leads to a drastic notch stress reduction in the dark spot (Fig. 32) shown in

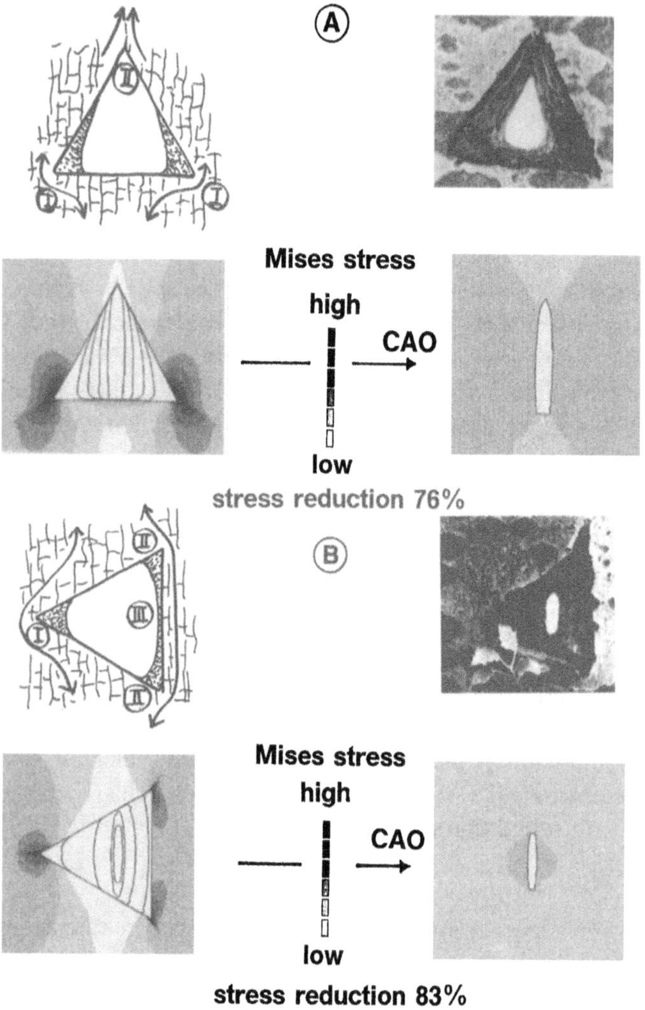

Fig. 32A,B. Computer simulation of the occlusion of a triangular wound in a tree. **A** Upright. **B** Flat

the area of the notch. Steps 2 to 5 are repeated until all notch stresses are eliminated or until growth must be stopped for design reasons.

Two to five iterations normally suffice for a successful optimization of engineering components. The number of repetitions depends on the quality and shape of the initial design proposal.

The advantages of CAO are evident:

- All that is needed is a commercially available FEM program. Since most of these programs are able to calculate thermal stresses, additional software for optimization is not required.
- Two- and three-dimensional components can be optimized without any need for specific problem-oriented solutions or tricks. It suffices simply to generate an FEM structure for the component one wishes to optimize.

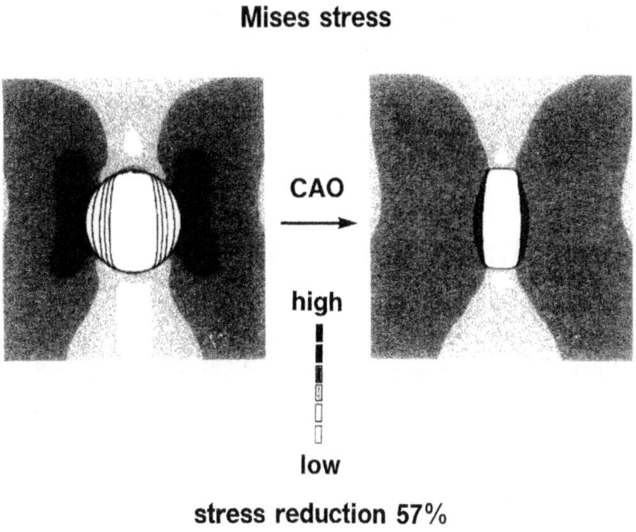

Fig. 33. Computer simulation of the occlusion of a circular wound in a tree

Two examples, a triangular notch and a circular hole, were chosen to elucidate the procedure. A triangular notch punched in a tree (Fig. 32) causes a force flow diversion that is most pronounced at the two lower edges (Fig. 32A). This pronounced force-flow diversion causes stress peaks (notch stresses), and indicates the areas of the greatest disturbance of an even stress distribution and also the points where material is attached to improve the situation. These notch stresses produce the greatest thermal expansions when transformed into temperatures. The growth of the selected structure is simulated on the basis of these greatest values. There is good agreement between theoretical simulation and the natural growth conditions. The triangular notch, which is turned by 90° (Fig. 32B), exhibits its highest notch stresses on the left side, while the upper and lower corners on the right divert the force flow more smoothly. Correspondingly, wound occlusion in the corner on the left-hand side is fastest, both when computer-simulated and when observed in nature. The circular hole (Fig. 33) reveals lateral stress peaks, and heals into the typical, natural vertical spindle described above. Evidently, wound healing is a mechanically controlled process whose theoretical prediction is achieved by applying the CAO method.

Since biological repair growth actually improves the "faulty designs" found in nature, a similar optimization of engineering components for example notch stress reduction by "growth", suggested itself. CAO, the component optimization method developed in accordance with the principles of biological growth, has been winning popularity throughout the German industry. An increasing number of companies and industries rely on the simplicity and efficiency of the method in finding lightweight, fatigue-resistant solutions. A detailed description of this successful method is found in Mattheck (1990, 1993).

6 Parallel Fiber and Force Flow Arrangement

6.1 The Principal Stress Trajectories are Shear-Free Lines

Wood that is subjected to tensile stresses along the grain exhibits no shear stresses between the fibers and no transverse tension, which could lead to fiber splitting (Fig. 34). If exposed, however, to tensile stresses at some angle to the grain, the fibers fall out of their optimum order, and high transverse tensile stresses and shear stresses occur, which threaten to split them. The risk of splitting is highest in the case of tension across the grain.

The method described below was developed at the Kernforschungszentrum, Karlsruhe (Kriechbaum et al. 1992) for determining optimum fiber arrangements even for complex geometries.

6.2 Computer-Aided Internal Optimization – CAIO

As described above, the fibers are arranged ideally when the axial force flow (principal stress trajectories) coincides with the orientation of the fibers. This optimum arrangement prevents shear stresses and transverse stresses from occurring in between.

Figure 35 shows a flow sheet of the CAIO method. CAIO can be applied based on any commercial FEM program. The optimum run of fibers is determined for a selected component, which is assumed to be subjected to defined load and service conditions. One may choose between an arbitrary initial fiber orientation or an isotropic structure without fibers. The orthotropic behavior of the selected structure varies locally and is described by the respective local system of coordinates. One axis of the system of coordinates coincides with the stiffest orientation, i.e., with the fiber orientation.

In a first step, the force flow is calculated for the nonoptimized fiber orientation. The local coordinate systems of the individual finite elements are turned until their axes coincide with the stiffest orientation (fiber direction) of the calculated force flow. The locally varying orthotropic distribution obtained provides the basis for another run. A second set of force-flow distributions is calculated, and the local orthotropic axes are aligned with the newly obtained distributions. The method is characterized by its rapid convergence. It takes only three to four iterations to eliminate shear stresses between the fibers and

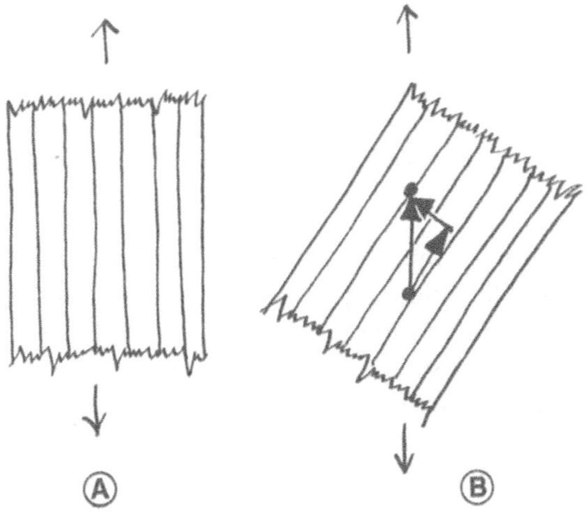

Fig. 34A,B. Force flow and direction of fibers. **A** When the fibers are aligned along the force flow, there is no tension transverse to them and no shear between them. **B** Transverse tensile stresses and shear stresses act between the fibers when loads are introduced diagonally

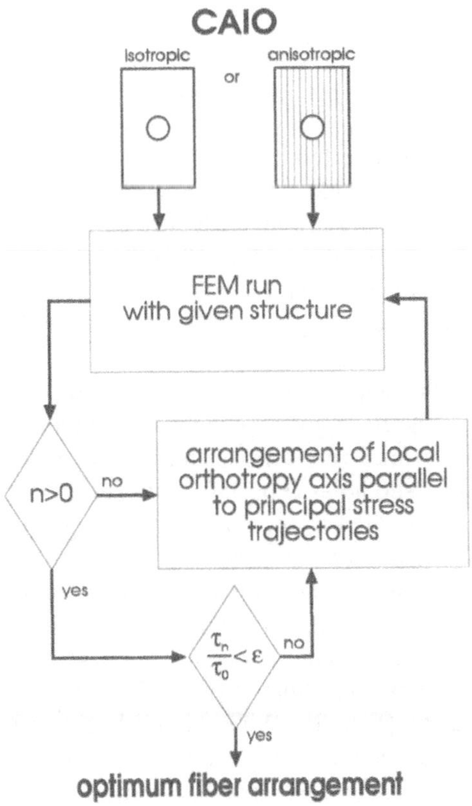

Fig. 35. The CAIO method determines the optimum run of fibers

obtain satisfactory final components whose fibers completely coincide with the prevailing force flow.

6.3 Spindle Shapes – Advantages and Problems

Figure 36 illustrates the optimization of the run of fibers around a circular hole. The diameter of the hole remains unchanged. Figure 36A shows the qualitative distribution of shear stresses before optimization, Fig. 36B represents the situation afterwards. High shear stresses surround the nonoptimized void, while the fibers that follow spindle-shaped contours are not subjected to shear.

A fiber composite structure with optimum run of fibers (C) was manufactured, and the resistances to tensile stresses of a nonoptimized prototype (D) and an optimized prototype (E) were tested by use of a tensile strength testing machine. The resistance to tensile stresses of the optimized component was found to be 1.5 times as high as that of the nonoptimized specimen.

The wood fibers are diverted around similarly spindle- or almond-shaped ray cross sections as is shown for *Acer platanoides* in Fig. 36F. These almond- or spindle-shaped cross sections divert the axial force flow smoothly and minimize the shear stresses between the fibers.

A smooth diversion of the fibers is also observed in the case of larger obstacles. A 15-mm hole was drilled across the branch of a poplar to study the process of wound occlusion (Fig. 37). The cambium above and below the hole receded, and the newly formed fibers were diverted, taking the spindle-shaped course. Wound healing does not progress continuously on account of the width of the hole. The tree reduces the shear stresses in the newly formed fibers by diverting them around the hole in alignment with the force flow. The load on the perforated tree rings below remains unchanged for a while but the bending stresses and shear stresses are reduced increasingly as wood is attached in the course of wound occlusion.

Evidently, the tree resorts to spindle-shaped "circumnavigation" to profit from the advantages discussed above. Figure 38 gives some more examples of spindle-shaped runs of fibers around various tree forms such as callus collars (Fig. 38A), healing wounds (Fig. 38B), deep holes (Fig. 38C), rays (Fig. 38D), the apertures of pits (Fig. 38E) or a parallel-to-the-grain cut through the membrane of a bordered pit (Fig. 38F). The rays may seem ideal on account of their spindle-shaped cross sections. However, they are weak points in the wood and cannot compete with a strictly uniaxial, parallel fiber orientation.

Two bicycle spokes make a useful model and metaphor due to axial tensile or compressive stresses that act in the ray-and-fiber profile (Fig. 39). The spokes represent the fibers next to a ray (Fig. 40). While lateral buckling is observed as one exposes the spokes to pressure, the ray is found to be torn in the transverse direction (rope, Fig. 40A). When the spokes are subjected to

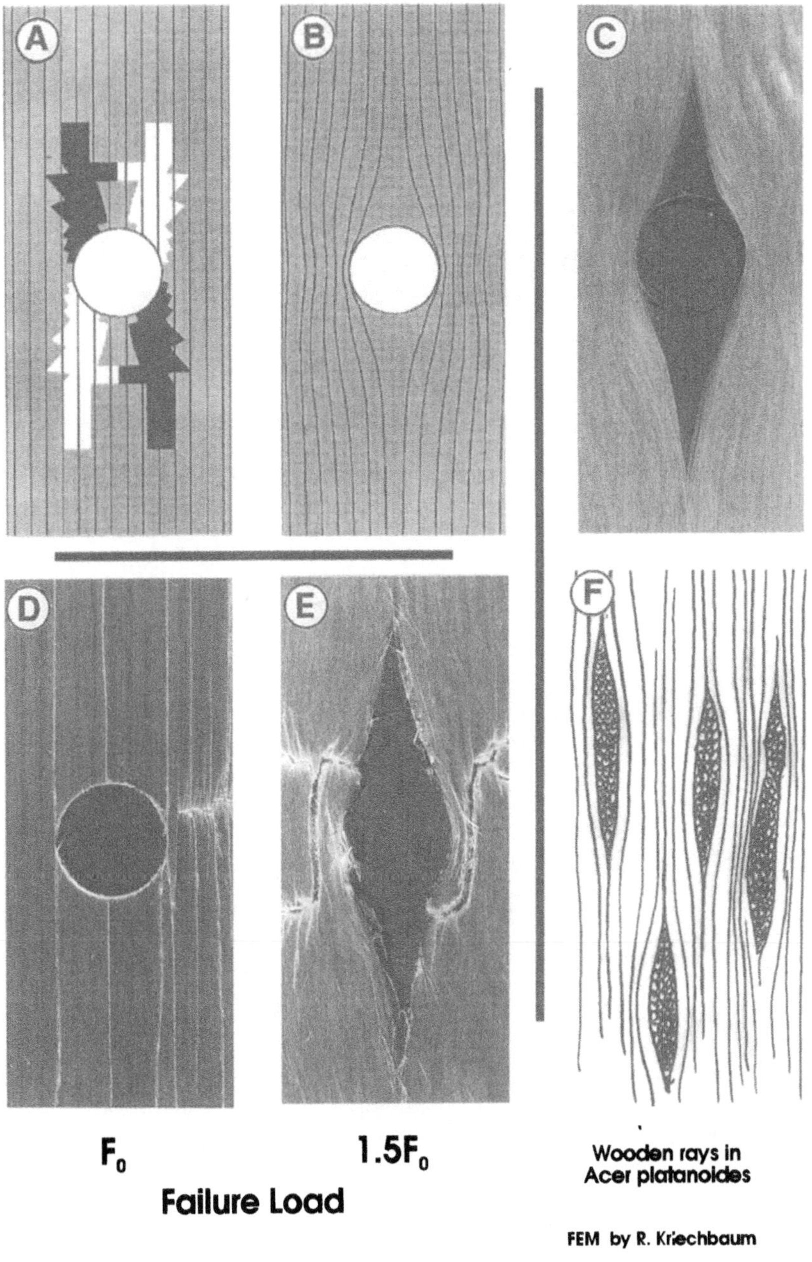

Fig. 36A–F. A spindle-type run of fibers around a circular hole reduces the shear stresses drastically. This spindle-type course resembles the run of the fibers around the rays

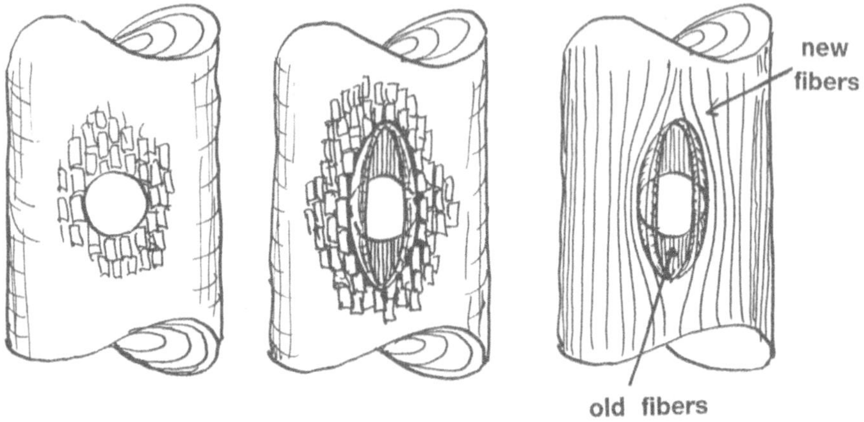

Fig. 37. A similar spindle-type course is found around a bore hole passing through the branch of a green poplar

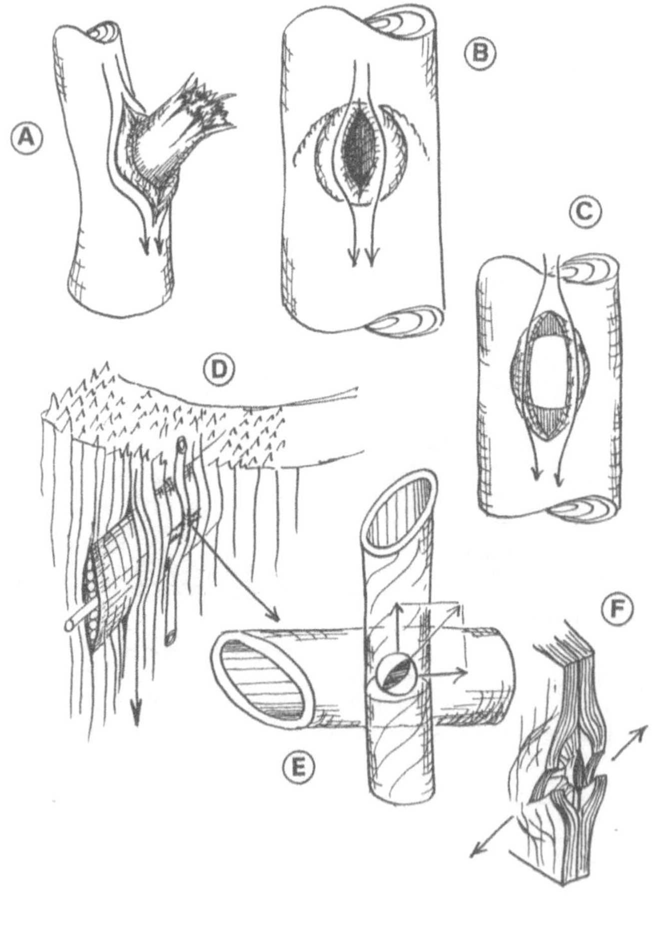

Fig. 38A–F. Three-dimensional spindles in the area of collars, wounds, rays, and bordered pits

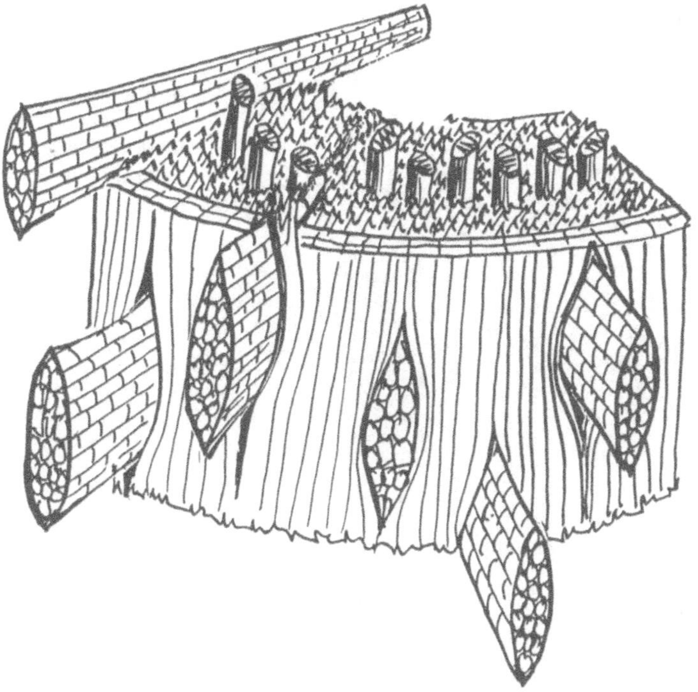

Fig. 39. Three-dimensional representation of the orthogonal ray-and-fiber structure

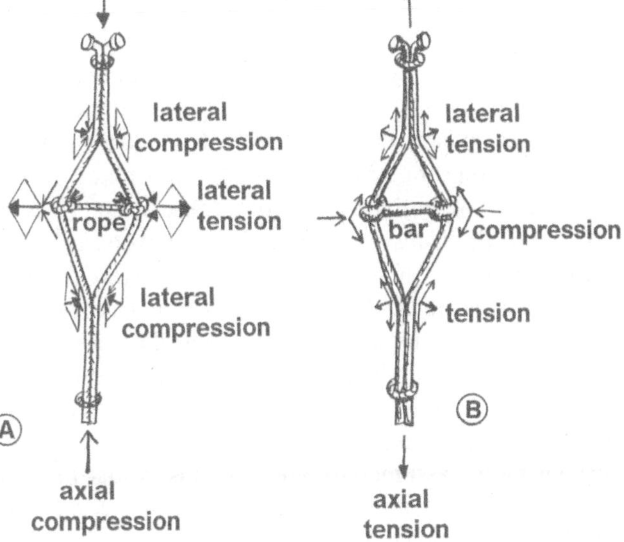

Fig. 40A,B. Two bicycle spokes demonstrate the failure of spindles under external loading along the grain due to unavoidable transverse stresses

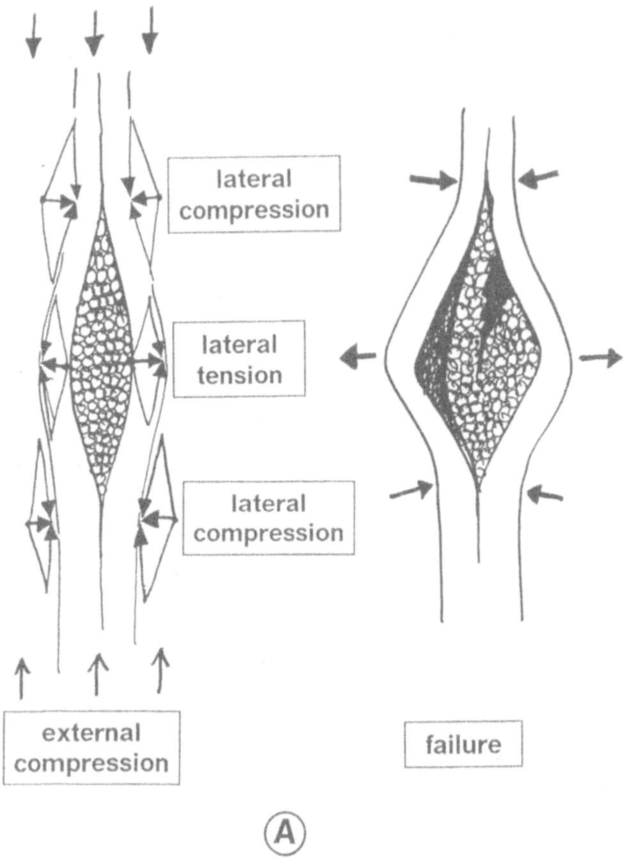

Fig. 41A,B. Failure caused by transverse stresses in the area of the rays. **A** Under external compressive loads. **B** Under external tensile loads

axial tension they tend to compress the enclosed ray transversely (strut, Fig. 40B) in the middle, but the spokes are pulled away laterally from each other below and above the ray.

Two types of failure mechanisms can be derived from this model for tension and for compression along the grain, i.e., along the spokes. In both cases, the rays have the effect of initial defects or "crack starters". Small longitudinal cracks develop as the transverse strength of the wood decreases (Fig. 41). Under axial compression (Fig. 41A) these cracks start from the center, which is the point of rupture of the rope (Fig. 40A). The ray either comes off the surrounding fibers or breaks in the vertical direction. Under axial tension (Fig. 41B) the fibers delaminate above and below the ray while the center of the ray is compressed transversely. Tearing cracks develop in the upper and lower corners of the ray cross section.

Similar failure mechanisms are assumed for knots because knots, like rays, are initial defects that are circumnavigated by the fibers following comparable spindle-shaped lines. Figure 42A shows the model of a branch joint (Shigo

Fig. 41B

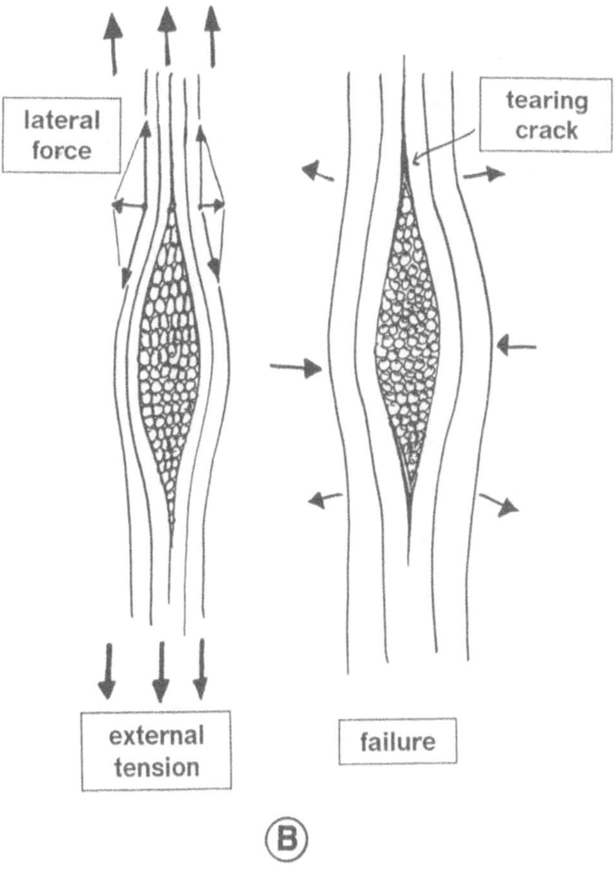

1990), which consists of a laminated composite of downstem branch fibers and stem fiber collars. The branch surrounded by the stem fibers may be considered the larger counterpart of the ray.

Under axial compressive loads (Fig. 42B), stem fiber delamination accompanied by lateral fiber buckling is observed in the area of the joint. Under axial tension (Fig. 42C) splitting occurs again above and below the branch. The seemingly harmless conglomerates of small lateral fissures may combine into large cracks, which can finally threaten parts of the tree or the entire biological component.

Axial loading is not the greatest threat to the ray environment. When the rays are pulled laterally, i.e., across the grain, their pointed edges act like crack tips that cause high notch stresses and increase the risk of splitting (Fig. 43). Rays that are subjected to tensile stresses in the direction of the tree rings (tangentially) thus turn into cracks. This phenomenon becomes evident in a dried slice of green wood with strange and curious tree rings (Fig. 44), a result of bygone injuries and wound occlusion processes. It is because of the orthogo-

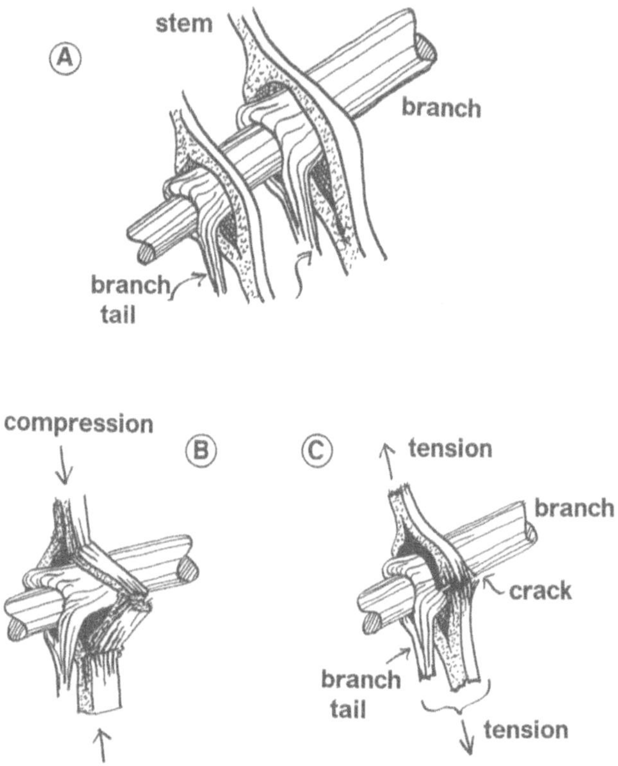

Fig. 42A–C. Failure near knots due to transverse forces caused by axial external loading. **A** Intact branch. **B** Failure under axial compression. **C** Failure under axial tension

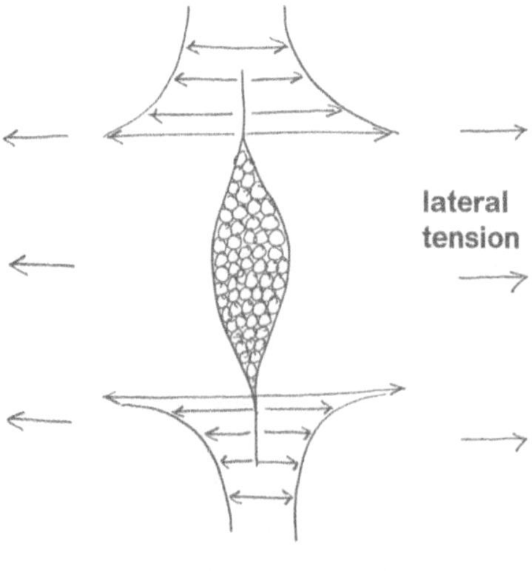

Fig. 43. Under transverse tensile loading in the circumferential direction, the rays have the effect of cracks

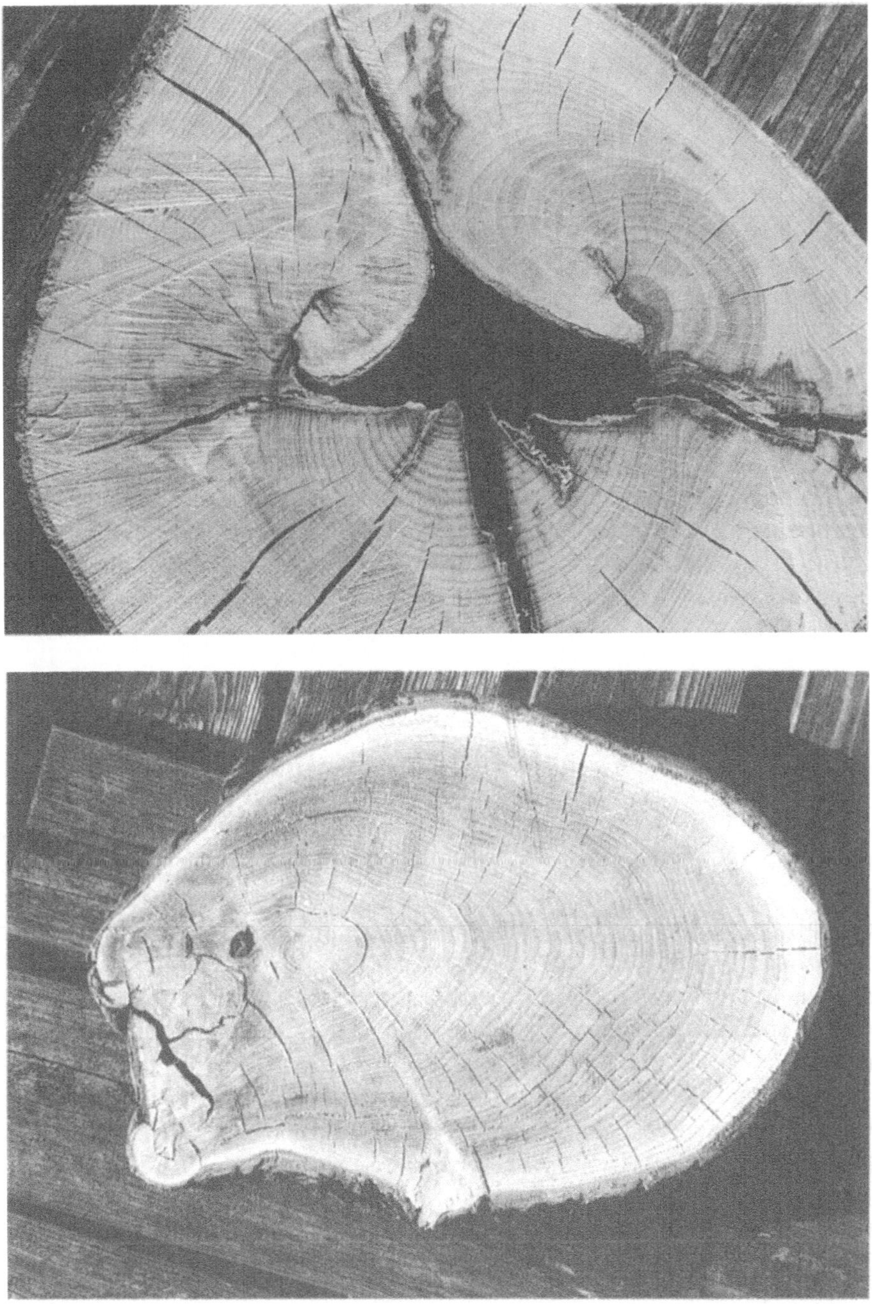

A

B

Fig. 44A,B. Drying slices of wood reveal the cracking characteristics of the rays: the cracks run right along the rays. **A** Beech tree with cracks caused by drying. **B** Oak tree with cracks caused by drying

nal system of the tree rings and rays at any point that the orientation of the rays in such slices is as exotic as the orientation of the tree rings. Since the rays are dormant cracks that are activated by tangential drying stresses, the appearance of the drying cracks is as curious as the arrangement of the rays. Naturally, regular slices with normal tree rings also crack along the rays. However, a convincing, unequivocal proof of the cracking characteristics of rays is their trick of pursuing even the strangest and most bizarre bends of the tree rings. A section on growth stresses later shows that the rays are weak points that the tree tries to control and fight by lateral compression, i.e., by reduction of the tangential tensile stresses that convert the rays into cracks.

6.4 Fiber Deviations

In this book, any tree whose grain does not run parallel with the trunk axis is defined as a tree of the spiral-grain type. Spiral grain may have different causes. The purely mechanical factors that induce spiral-grain formation are discussed below. A similar but perhaps less rigorous definition is given by Harris (1989).

6.4.1 Mechanically Controlled Spiral Grain

A cylindrical bar that is twisted round its longitudinal axis develops a spiral force flow. Figure 45 shows the principal tensile stress trajectories and the fibers arranged along them. The principal compressive stresses act perpendicularly to the orientation of fibers. They compress the fibers, reducing the risk of splitting and the shear in between. Spiral grain caused by external mechanical torsional loading reveals that the tree adapts to the given load case. It specializes, so to speak, in torsional loading in one particular sense of rotation.

Such one-sided, fixed fiber orientations are ideal as regards the respective loading, but involve a certain risk of failure under changed load conditions (see sections below). In the case of unidirectional torsional loading, torsion-adapted spiral grain minimizes the shear between the fibers and reduces the risk of delamination by restricting the trend towards mutual lateral fiber compression. The run of the fibers and the force flow are inclined towards the trunk axis (Fig. 45).

Only some particularly exotic samples of the countless manifestations of spiral grain are described here. The harp conifer shown in Fig. 46 developed due to the slanted growth of an almost mature trunk, which had grown too thick to be able to straighten under the influence of negative gravitropism. The thinner and more flexible lateral branches that then grew upright remind one

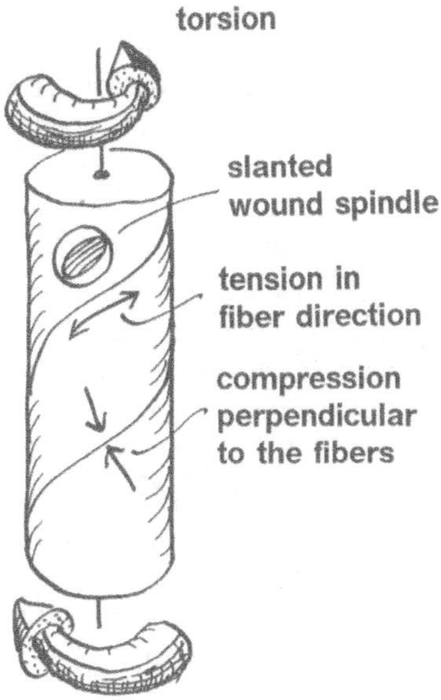

Fig. 45. Under torsional loading the principal stress trajectories, i.e., the force flow, run along spiral lines

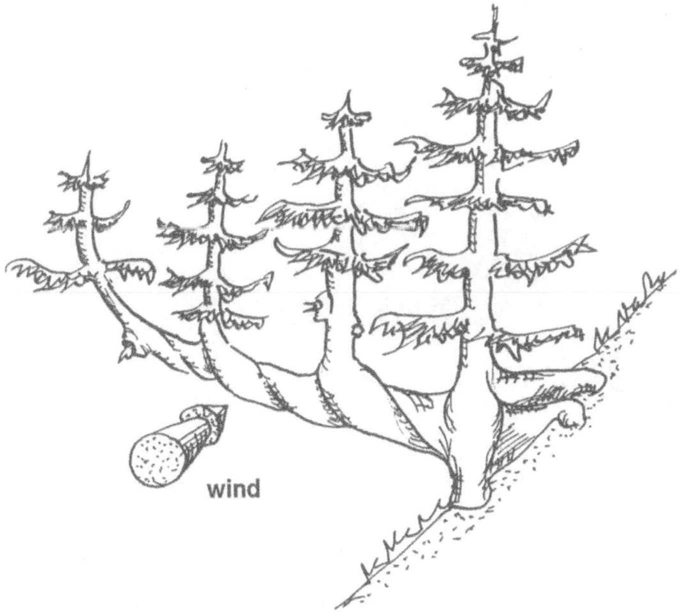

Fig. 46. Harp tree with spiral grain (enlarged scale) caused by torsion

of the strings of a harp. Under unilateral wind loads (Fig. 46), these string-type vertical branches are exposed to bending, while the stem is subjected to bending and torsion. Such harp trees are often found to exhibit enormous stem diameters in the areas of the lateral branch joints because the branch-bending loads are turned into local torsional loads on the stem.

Figure 47 represents a pendulum beech which was exposed to changing wind loads subjecting the centric vertical stem to pure bending in two directions. This load history can be inferred from the differing windings of the fibers on both sides of the stem. Each side is wound so that the fibers of one side of the horizontal stem are subjected to tension whenever the pendulum is exposed to bending in one or the other direction. The fibers are wound like spirals on account of the torsional load.

Excessive reaction wood formation explains the spiral grain of the slanted conifers shown in Fig. 48. The growth of reaction wood (compression wood) on the lower sides of conifers is known to cause these sides to lengthen. A considerable difference in length (Fig. 48A) would show if one separated the upper from the lower part. The tree bends upward (Fig. 48B) because the upper side and the lower side adhere to each other, while the lower side

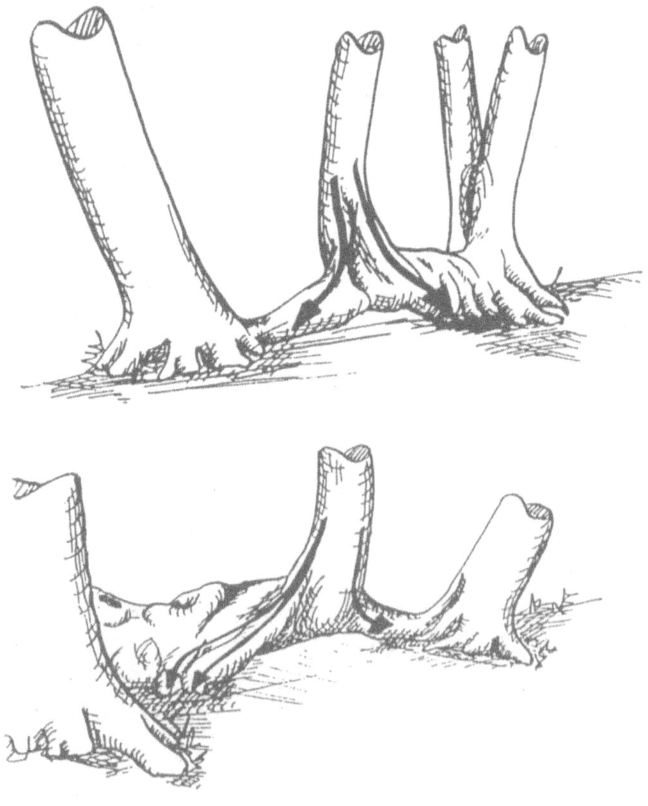

Fig. 47. Harp tree pendulum stem with bilateral spiral grain. Since the spirals run in the opposite direction, the stem can absorb vibrations without shear stresses in either direction

expands due to compression wood formation. Sometimes, upward bending seems insufficient to compensate for this enormous difference in length, and the tree starts to twist the longer lower side round the short upper side (Fig. 48C). Through this twisting, the former lower side becomes the upper side. Saw cuts of such stems show sickle-shaped reaction wood formations, characterized by lateral tree ring offsets. Obviously, this mechanism helps to balance the difference in length of the two halves of the tree.

Local spiral-grain formation at the foot of the trunk may be observed when a carrying root is lost or fails (Fig. 49). When the lateral root is intact, the downstem force flow enters the soil, running along the root fibers and passing a labyrinth of ramifications. When the root is cut or when it is destroyed by

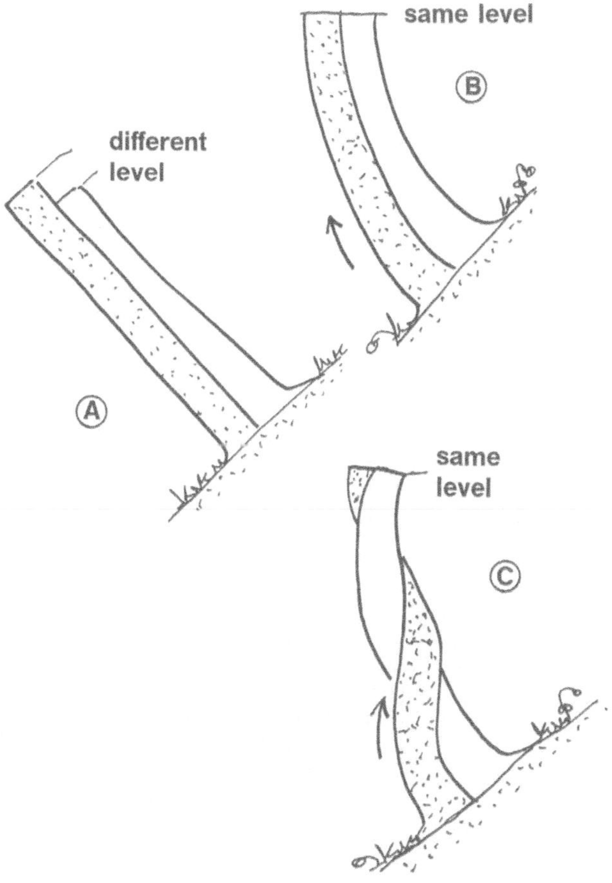

Fig. 48A–C. Compression wood and spiral grain. **A** If the two halves yielded, their lengths would differ. **B** Shear-resistant coupling makes the two halves bend and grow upright. **C** When self-bending does not suffice to compensate the different levels, the lower (longer) half of the stem winds around the upper (shorter) half, thus causing the formation of spiral grain

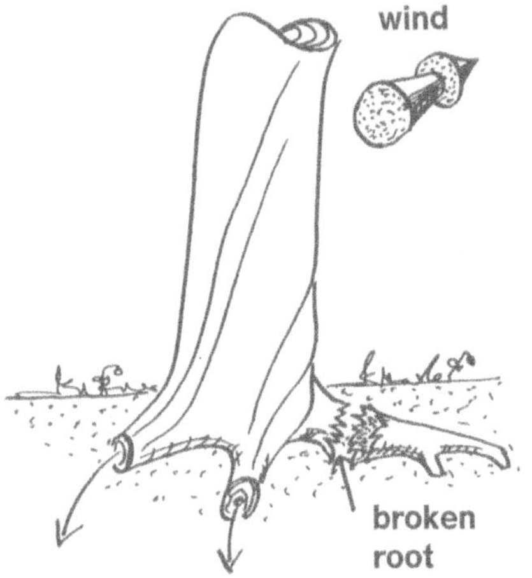

Fig. 49. Spiral grain at the butt of a stem due to the loss of a big root

Fig. 50. Force-flow diversion by slanting notches is similar to that caused by strangling wires or vine shoots. The new fibers run along the diverted force flow, and lead to the formation of spiral grain

fungi, the force flow must find its way through the neighboring roots, and new connecting fibers develop in the course of the years. In accordance with the spiral-grain definition given by Harris (1989), the fibers are inclined relative to the trunk axis as they pass from the trunk above the defective root into the neighboring load-carrying roots.

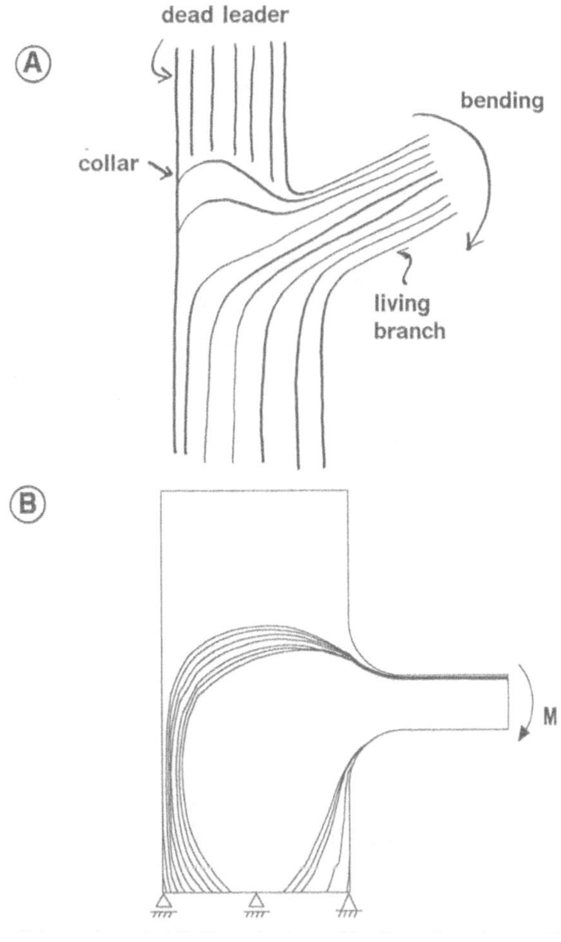

Fig. 51A,B. After the death of the leading shoot, the fibers are arranged along the force flow introduced by the branch. **A** Direction of fibers after Neeff (1914). **B** CAIO prediction of the orientation of fibers. (Calculated by R. Kriechbaum)

Local spiral grain may also develop in the presence of slanted notches or vine tendrils (Harris 1989); Fig. 50 shows an example. Such manifestations of spiral grain are not necessarily permanent. The fibers may resume their axial orientation to bridge the wound after occlusion, provided that there are no other stimuli to induce spiral grain. Additional shear is observed, however, when the newly formed fibers cross due to their abrupt reorientation.

Changed force-flow and sap-flow conditions after decapitation of a leading shoot may lead to spiral-grain formation in the area of the adjacent branch joint (Harris 1989). Figure 51A illustrates the collar-type run of fibers round the dead leading shoot. The fibers are not necessarily symmetrically arranged. Drastic differences are revealed if one compares the joints of dead leading shoots with the joints of living stems (Shigo 1990). While drop-shaped branch fiber formations are found between the collars of the vital stem (Fig. 42), the collar round the remnant of the dead leading shoot (Fig. 51A) is formed by the branch itself. The appropriateness of the course of the force flow in the latter

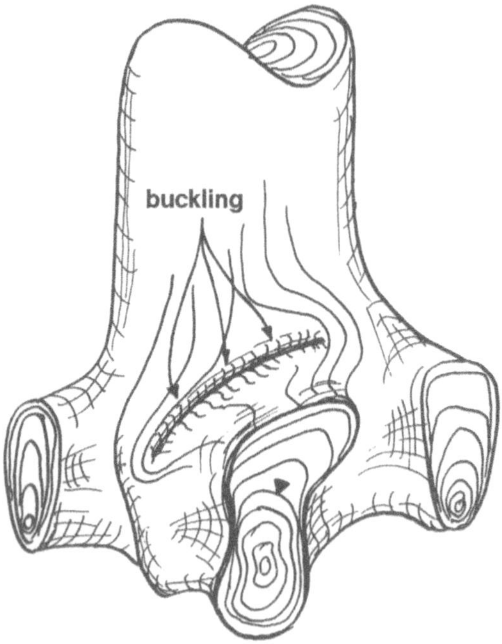

Fig. 52. Force-flow diversion and spiral-grain formation may be caused by local fiber buckling

case is corroborated by the approximate CAIO determination of the optimum fiber orientation in a two-dimensional model (Fig. 51B). It is plausible that the fibers are inclined towards the former stem axis. Since neither a significant force flow nor a sap flow comes from the dead leading shoot, there is no need for vertical-fiber formation. Vertical fibers would be out of place and would turn such trees into faulty designs.

The orientation of fibers at the lower clamping in the calculation model (Fig. 51B) is out of line because the mechanical boundary conditions do not exactly correspond to the real conditions. Figure 52 shows that spiral grain may also be caused by local fiber buckling in the stem-to-root transition region. No stresses occur in the buckled area, and the force flow must be diverted on both sides. The tree rings in the root cross section indicate the year of the accident (black triangle). The distribution of stresses in the root is changed, and the change in the stress distribution causes a change in the thickness of the tree rings that were added after the accident. The root cross section is an authentic, dependable record of the tree's load history (Mattheck, Breloer 1993).

Inclined orientations of fibers may develop when the force flow winds its way through between two close-set branches (Fig. 53). It is due to the secondary thickness growth of the branches that the fibers are squeezed or compressed as they pass the bottleneck. External loading plays a secondary role as the branches fight and "elbow" for their increase in diameter. It may be misleading to speak of spiral grain when referring to such formations. Spiral

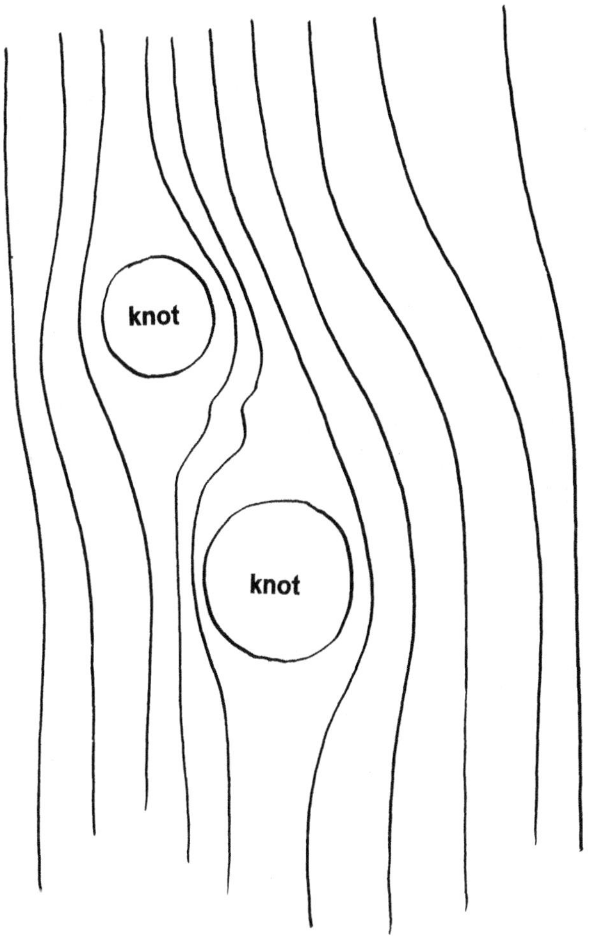

Fig. 53. Local spiral-grain formation may result from vertical and lateral branch joint offsets

grain is, however, defined here as any run of fibers that deviates from the stem axis (Fig. 53). It is certainly difficult to differentiate clearly between spiral grain and normal fiber orientations, because any branch joint or any diversion of the fibers around rays is a spiral-grain phenomenon in the sense of Harris' definition.

Any discussion of the diverse spiral-grain manifestations would be incomplete if one restricted it to the above examples of mechanically controlled growth, while forgetting the hereditary forms that develop irrespective of their mechanical advantages for survival.

6.4.2 Genetically Programmed Spiral Grain

A study by Kubler (1991) describes the biological aspects of spiral grain and discusses some of its original causes in addition to the purely genetic disposition. Kubler's study emphasizes the biological advantages of spiral grain. Spiral grain essentially contributes to a more constant circumferential distribution of nutrients in the case of branch losses, and of water in the case of root losses. It stands to reason that spiral grain is a helpful mechanism regarding the elaborate, ingenious windings supporting the trunk. The advantages of spiral grain are accompanied by certain disadvantages, described in a discussion of the risks of spiral grain.

6.4.3 The Risks of Spiral Grain

Any mechanical component that is optimized for a certain load case is particularly well designed and prepared for the strains and stresses of that specific load. The engineer puts up with an exclusive load case specialization while profiting from the advantages of the tailor-made structures that he develops.

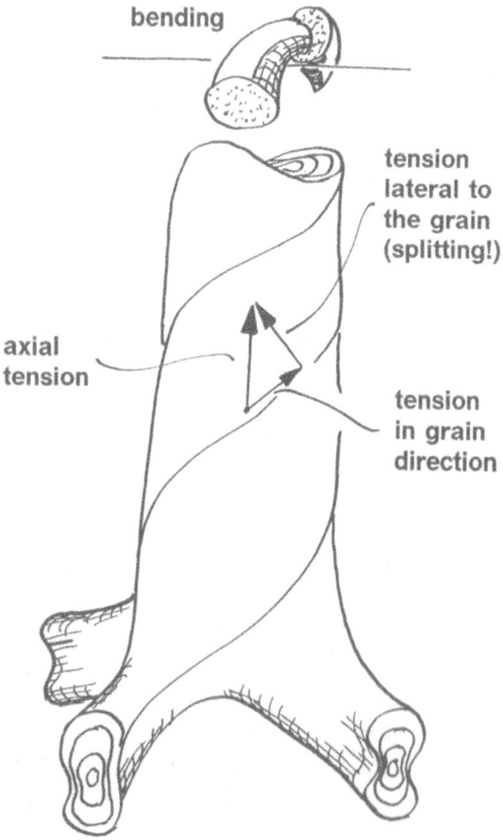

Fig. 54. Stems with spiral grain are exposed to shear loads and tension perpendicular to the grain when subjected to pure bending stresses

The drawbacks of specialization are particularly evident in the case of spiral grain. Figure 54 shows a spiral-grain tree subjected to pure bending because of its largely symmetrical crown. Unlike the winding force flow observed under torsional loads (Fig. 45), the force flow resulting from pure bending takes a vertical course. This vertical force flow can be resolved into vectors (Fig. 54) to obtain a force along and a force perpendicular to the grain. The perpendicular force may induce decay-inviting torsional cracks under transverse overloads. Trees whose spiral grain is optimized for torsion can succumb to pure bending, and thus turn out to be of faulty design.

Trees whose fiber orientation is unsuited for such loading may counteract failure by improving the tensile strength of their middle lamella. This speculative load-adapted strength distribution has been observed in trees within the framework of other studies (Dietrich and Mattheck 1994).

A faulty design will deteriorate if, in addition to bending, one admits torsion acting in the opposite direction of the spiral grain (Fig. 55B). The advantageous load directions in Figs. 45 and 55A are reversed, while the tenacious fibers are exposed in the longitudinal direction to compression and in the transverse direction to tension. Spiral longitudinal cracks, which sur-

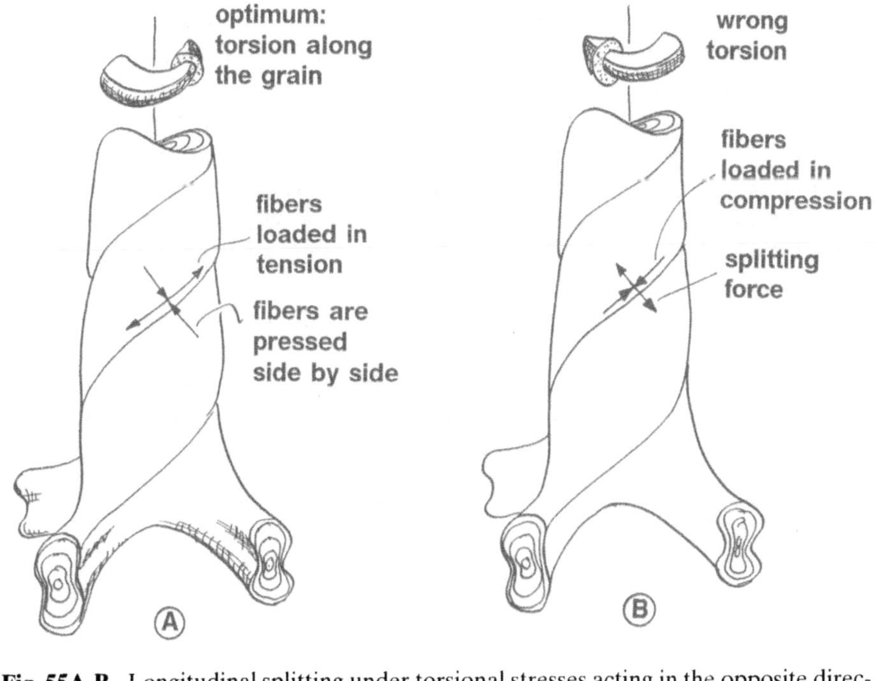

Fig. 55A,B. Longitudinal splitting under torsional stresses acting in the opposite direction of the spiral grain. **A** In the right sense of rotation the fibers are drawn longitudinally and are compressed transversely. **B** In the case of torsion in the opposite direction of the spiral grain, the fibers are compressed longitudinally and are drawn apart transversely

round the stem following the direction of the spiral grain, are bound to develop. Spiral-grain trees can be compared to ropes. The fibers are exposed to longitudinal tension and are compressed transversely when the rope is twisted in the direction of its windings (Fig. 55A). The same self-compacting effect is observed in the case of trees that are twisted in the direction of the spiral grain. The rope shown in Fig. 55B unwinds and comes undone due to opposite loading.

The above phenomenon is also evident in reinforced concrete. This composite material is supposed to resist tension (steel bars) and bear pressure (concrete). Turning the concrete slab upside down, the steel bars are compressed and are turned into struts, while the pressure-resistant concrete easily ruptures under the tensile load.

Any optimization is a design load specialization. Optimization may become an act of sabotage if the design load turns out to disagree with the service load. This applies to all mechanical components, including trees.

The ignorance of this simple truth has caused many a notable, historic tree to fail due to changes in the wind load after, for example, the construction of new buildings nearby. A 400-year-old oak tree with spiral grain failed due to inadequate torsional loading (Fig. 55B) after its habitual sense of torsional rotation had been reversed by the failure of adjacent trees in a storm. The direction of torsional rotation is also changed through crown reductions, which reverse the symmetry of the canopy. Such wrong-headed tree care measures may convert parks or avenues into dangerous places lined or covered with faulty trees weakened by torsional cracks throughout. Trees with natural spiral grain, e.g., horse chestnut, often become the victims of unilateral crown reductions and reversed torsion. Hereditary spiral grain is believed to be more susceptible than mechanically controlled spiral grain to torsional-crack formation. This tendency of the hereditary types may be due to their higher risks of adverse loading. The optimum load-adapted designs of trees with mechanically stimulated spiral grain are not disturbed, except by drastic changes in the wind direction or through radical shifts in the symmetry of their crowns.

Evidently, the outer fibers are aligned with the force flow to the advantage of the tree, provided there are no contrary genetic factors. In spite of the risks discussed above, the biological and mechanical advantages of spiral grain predominate, and have been justifying the survival of spiral-grain formations throughout evolution.

Summary

- The orientation of the outer fibers is controlled mechanically and follows the force flow unless hindered by genetic programming. The fibers in the center of the tree are no longer aligned with the force flow.

- The arrangement of the fibers along the force flow reduces the risk of splitting by minimizing the shear stresses between the fibers and the transverse tensile stresses.
- The fibers are diverted a round obstacles that are in the way of the force flow. The fibers follow the contours of spindles as they circumnavigate the obstacles.
- The diversion of the fibers explains the macroscopic (e.g., wound spindles, stem fibers surrounding branches) and the microscopic (e.g., ray cross sections, bordered-pit formations) orientation of fibers.
- Spindles may fail when subjected to transverse tensile stresses. They may develop into cracks or take on the properties of cracks.
- Mechanically controlled spiral grain follows the principal tensile stress trajectories.
- Genetically programmed spiral grain ignores the run of the force flow with respect to the grain.
- In the presence of spiral grain, nutrients and water are distributed more evenly in the case of tree crown or root losses.
- Torsion in the opposite direction of the spiral grain causes torsional cracks, i.e., longitudinal splits along the spiral grain.

7 Tree Rings and Force Flow

7.1 Longitudinal Tree Ring Arrangement

The ground tissue of the tree rings consists of fibers that cannot escape and desert to the neighboring formations that developed at an earlier or later stage. This being so, the longitudinal tree ring formations obey the same fiber-and-force-flow concurrence rules as the manifestations discussed above for the fibers. When the run of the tree rings is adapted to the force flow, the shear between adjacent structures is minimized and the transverse splitting forces are reduced. This shear stress and transverse tensile stress relief is a useful mechanism if one considers that the latewood/earlywood boundaries are welds that do not consist of additional material attached continuously in the course of time.

Figure 56A shows an example of load-adapted tree ring arrangement. The parallel cut through a tree that had contacted a stone reveals that the tree surrounded the stone to enlarge its contact surface and reduce the high contact stresses. The tree rings align perpendicularly to the surface of the contacting stone.

7.2 Tangential Tree Ring Arrangement

The significance of tangential growth stresses (see Chap. 8) becomes evident when considering their mechanical force and their relevance to the resistance of the tree to cleaving. In addition to these tangential loads on the tree rings, one finds external contact stresses resulting from secondary thickness growth. A small cylinder that is tightly screwed to a tree (Fig. 56B) impedes the radial growth of the tree because the screws do not extend and do not yield under tensile stresses. The tree presses against the cylinder and is stopped in the effort of expanding. Numerous saw cuts revealed that tree rings align perpendicularly to the contact surface (A-A′) whenever the tree contacts a foreign body. The tree parts growing around the foreign body meet. They weld when the tree rings coming from both sides join without showing kinks or bends. The enclosed bark is penetrated at that stage, and a tree ring is formed that surrounds the foreign body. Trees can integrate objects such as traffic signs, pipes, or fences, by forming such enveloping tree rings (Mattheck 1993).

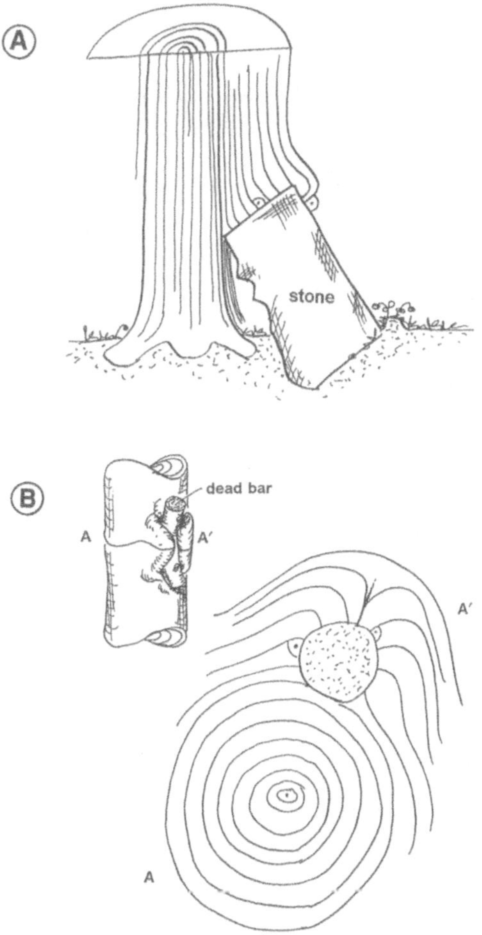

Fig. 56A,B. Tree ring arrangement perpendicular to the contact surface minimizes shear between the tree rings. **A** Longitudinal direction. **B** Transverse direction

The tree rings are aligned perpendicularly to the contact surface at any point, regardless of the complexity of the contacting foreign body. In the case of welded trees or tree parts (Mattheck 1991), each of the contacting partners aligns its new tree ring perpendicularly to the contact surface. As in the case of tree growth around foreign bodies, fusion of trees or of tree parts occurs when the tree rings meet to form one continuous, kink-free line.

Summary

- The tree rings consist of fibers and follow the axial direction of the fibers, which is normally adapted to the force flow.

- The tree rings follow the force flow also tangentially. The tree rings are aligned perpendicularly to any contact surface.
- The adaptation of the tree rings to the force flow minimizes critical shear stresses and transverse tensile stresses in the area of the tree ring boundaries.

8 Growth Stresses

8.1 Qualitative Description

Growth stresses were first investigated and described by Jacobs (1938). Kubler (1987) examined the growth stress phenomenon and established the simple theory discussed here.

Figure 57 illustrates and proves the existence of growth stresses in living trees in a qualitative but effective way. Circumferential growth stresses act as compressive stresses on the tree surface while cell contraction causes tension in the longitudinal direction. While radial saw cuts are compressed (Fig. 57A), longitudinal ones are bent apart at their free ends (Fig. 57B). The advantages and risks of such stress distributions are discussed later.

Reaction wood formation by leaning trees may be considered the counterpart of growth stress development in straight trees. A brief introduction to the mechanisms of reaction wood formation is given to facilitate access to the origin and causes of growth stresses. Figure 58A shows the effect of compression wood in gymnosperms (Timell 1986). Compression wood is characterized by its property of expanding longitudinally. It develops at the lower sides of leaning trees and pushes them upward into the vertical position. The walls of the tracheids of compression wood are thicker and richer in lignin than those of normal wood cells because of the higher risk of buckling. Also, the cells are shorter than in normal wood. A short, stiff, thick-walled tube that is exposed to axial compression does not bend or buckle as easily as a slender, flexible, thin-walled tube.

The negative gravitropism mechanism works differently in angiosperms. Tension wood develops at the upper sides of such trees. The tension wood contracts and acts like a rope pulling the leaning tree into position. The cells of tension wood fibers, which are subjected to tensile stresses, are not as rich in lignin but they contain more cellulose than normal fibers. A lignin-free so-called gelatinous layer forms the inner cell wall.

The growth stresses acting in trees are strong enough to bend mighty trunks and change their shape in the course of time. The mechanisms behind this curious phenomenon are easily understood if one neglects the biological aspects and takes a closer look at the pneumatic principle. The walls of the rubber tires of the cars and trucks that we drive are soft and thin, yet strong and durable enough to support and carry the heaviest loads and weights. They do not buckle and fail, because they are prestressed for tensile stresses by the pressure inside the inflated tire. There is no doubt that the

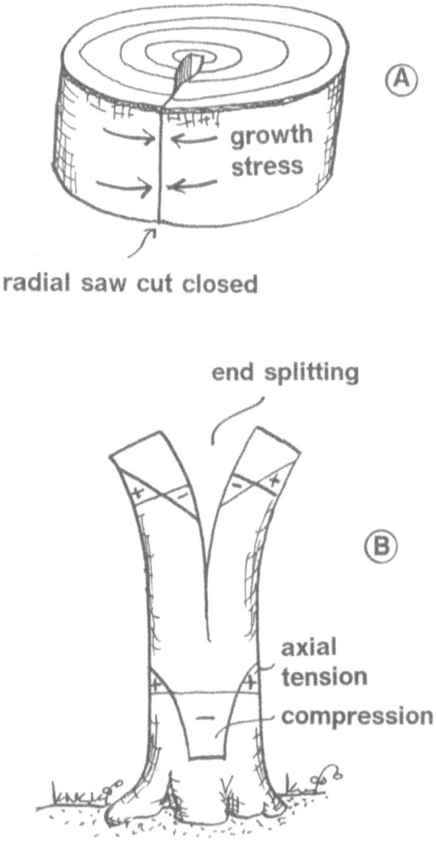

Fig. 57A,B. Qualitative distribution of growth stresses and growth stress detection. **A** The compressive stresses acting on the surface of the tree in the circumferential (tangential) direction compress a radial saw cut. **B** In the axial direction, tensile stresses act on the surface while compressive stresses act in the center of the stem. A diametrical cut causes the two halves of the stem to bend apart

filigreed walls of the cells of wood are subject to the same pneumatic principle during ontogeny.

The pneumatic principle refers to the cell inflation mechanism, but gives no explanation of the differences between axially expanding compression wood and axially contracting tension wood. These opposite effects of the equally inflated wood cells must be due to yet another feature of nature. Figure 59 gives a plausible explanation for compression wood. The fibrils in the walls of compression wood cells are known to be arranged quasi-transversely to the cell axis (Fig. 59A). The lateral expansion of the cell is impeded when the pressure inside increases. This constraint is missing in the longitudinal direction of the cells. The longitudinal extension of the cells under the influence of internal pressure increases may explain the effect of compression wood. Lignification after extension may help to keep the growth stress level high.

The fibrils also profit from their transverse orientation in the case of axial compressive loading. The typical fibril arrangement eliminates the risk of buckling and avoids failure.

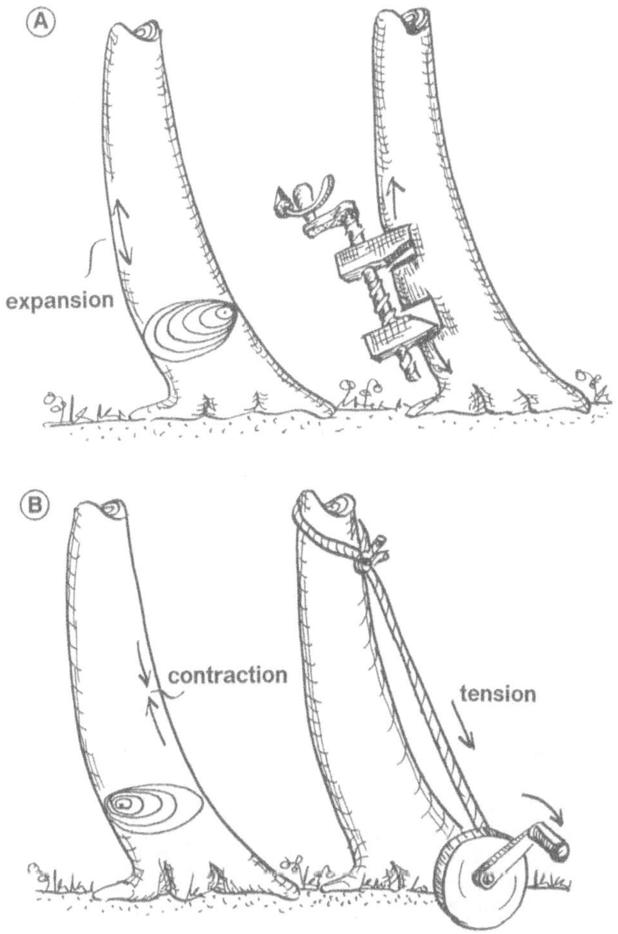

Fig. 58A,B. The effect of reaction wood in a slanted tree. **A** The compression wood of gymnosperms expands in the axial direction and pushes the tree into the vertical position. **B** The tension wood of angiosperms shrinks in the axial direction and pulls the tree into the vertical position

An ingenious trick of nature converts the axial compression of the tree into tension acting on the resistant, tenacious fibrils. A schematic sketch (Fig. 59B) shows the almost exclusive lateral expansion of a balloon which is bound with ropes. Such cross-windings would be fatal around tension wood cells under axial tensile loading. The windings would delaminate because of their poor adhesion, and would uncoil like the spirals of a spring. Since the fibrils are designed for high tensile loading, one finds them arranged in the direction of tension in tension wood (Fig. 60). Inflated tension wood cells assume a somewhat rounder shape, and contract longitudinally while expanding transversely. This explains both the regular growth stresses in straight trees and the formation of tension wood at the upper sides of leaning trees. In accordance with this explanation, tension wood is a particular growth stress variety which is promoted by a special cell anatomy. One might imagine that the gelatinous layer maintains the shape of the tension wood cell like pressurized air maintains the shape of an automobile tire.

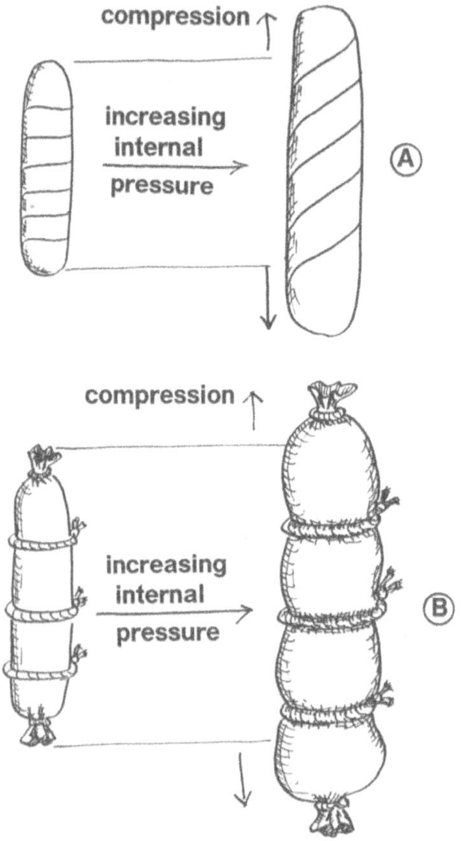

Fig. 59A,B. Compression wood fibrils are wound transversely to the cell axis. Under internal cell pressure, these fibrils expand in the axial direction, and induce longitudinal pressure. **A** Real situation; **B** simplified model

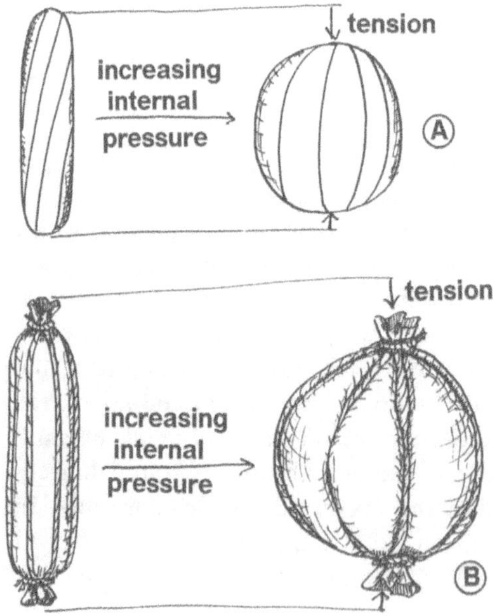

Fig. 60A,B. Tension wood fibrils are wound parallel to the cell axis. Their axial contraction under internal cell pressure effects longitudinal tension and tangential compression. **A** Real situation; **B** simplified model

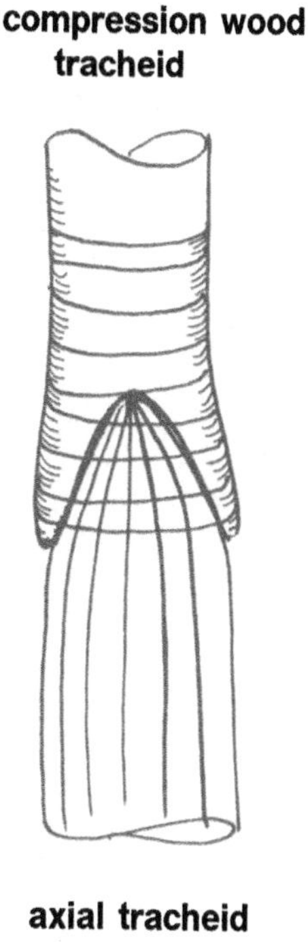

Fig. 61. A conceivable plug-in connection between compression wood tracheids and longitudinal tracheids

In speculating how the axial expansion of the compression wood cells is transmitted to the surrounding wood, there are some remarks in Wagenführ (1989) that maintain that compression wood cells are provided with bifurcated ends which reach into the longitudinal tracheids. "Bifurcated" is assumed to refer to two-dimensional sections and should be replaced by the more appropriate term "conical". One can speculate about a "plug-in connection" between the conically hollow compression wood tracheids and the conically convex longitudinal tracheids (Fig. 61). Such a connection would avoid delamination because the fibrils of the compression wood tracheids would be subjected to tension and the ends of the fibrils of the longitudinal tracheids would adhere to one another. This problem has not been investigated yet by the authors. The explanation which is given sounds plausible, but remains hypothetical until supporting findings can be provided.

The above-mentioned theoretical description of growth stresses in trees by Kubler (1987) gives three equations for the normal stresses:

longitudinal normal stresses

$$\sigma_Z = \hat{\sigma}_Z(1 + 2 \ln r/R), \qquad (11)$$

tangential (circumferential) normal stresses

$$\sigma_\theta = \hat{\sigma}_\theta(1 + \ln r/R), \qquad (12)$$

radial normal stresses

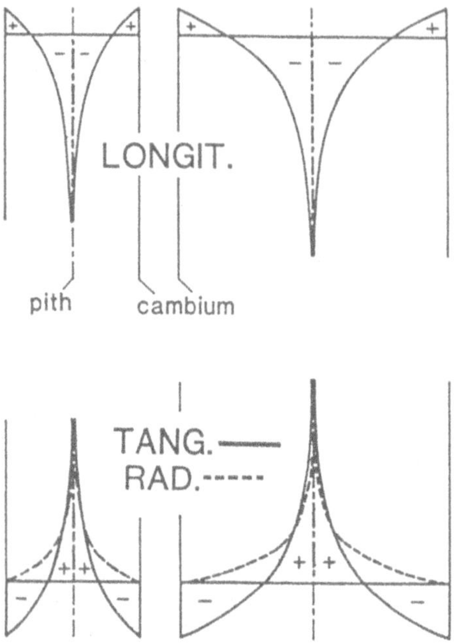

Fig. 62. Qualitative distribution of axial, radial, and tangential (circumferential) growth stresses after Kubler (1987)

$$\sigma_r = \hat{\sigma}_\theta \ln r/R, \quad (13)$$

whose quantitative distributions are plotted in Fig. 62. These three stress components and the effects they have on trees are discussed below.

8.2 Longitudinal Stresses

Kubler (1987) gave a very detailed description of the advantages of growth stresses in the longitudinal (axial) direction. The high risk of fiber buckling reveals that the compressive strength of wood is only about half as great as its tensile strength. Trees could resist much higher stresses if they had support in resisting the hazardous compressive stresses on their lee side. This assistance is, in fact, given by the axial growth stresses that act in them. Figure 63 reveals that the tree surface is prestressed by tension while the interior is prestressed by compression. As is evident, the marginal areas of the tree cross section are prestressed by tension. This prestressing minimizes the hazardous compressive stresses on the lee side, but increases the less critical tensile stresses on the windward side.

The resulting stress distribution, adapted to the strength of wood, would lead to failure due to tension and failure due to compression at the same time in the ideal case, for example, if springtime growth stresses were not higher than the growth stresses produced in winter. Growth stresses are a typical

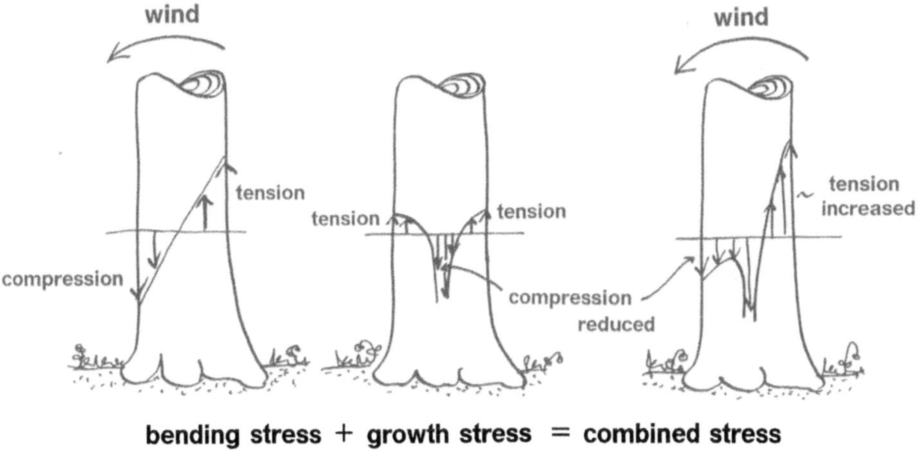

Fig. 63. The axial tensile stresses on the surface of the tree minimize the critical compressive stresses induced by the wind-bending load

characteristic of the green tree. They are weaker in the intervals between the vegetation periods. The reduced sail areas of the broadleaf trees, made leafless by winter, help to limit the risks involved in periodic growth stress stagnation.

While the positive influence of the longitudinal growth stresses has been known for years (Kubler 1987), the advantages for the wood of tangential growth stresses in preventing failure have not yet, to the authors' knowledge, been evaluated.

8.3 Lateral Stresses

The radial stresses shown in Fig. 62 are not discussed here. The radial-stress component is not as relevant as the longitudinal and tangential stress components because radial surface constraints are completely missing.

The fundamental importance of tangential growth stresses is evident considering their relevance to the safety of trees. The hazards of decreasing tangential stresses are familiar to anyone who knows the serious risks and consequences of the summer drop of branches.

The underlying processes are understood if one reconsiders the spindle-shaped ray cross sections described in Chapter 6.3. The peculiar fiber-diverting shapes of the cross sections reduce the risk of fiber buckling under axial compressive loads in the area of the rays. Tangential tension, however, is found to convert the rays into cracks (Fig. 43). Drying slices of wood show cracks along the rays (Fig. 44). The rays are potential cracks that develop into real and active cracks under tangential tensile loading.

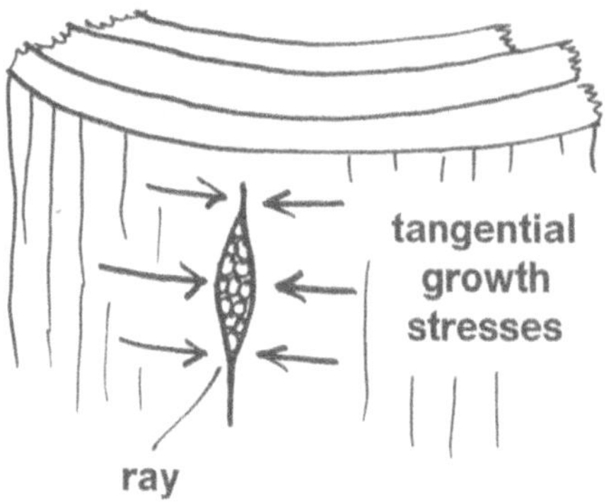

Fig. 64. The tangential compressive stresses on the surface of the tree compress the rays transversely

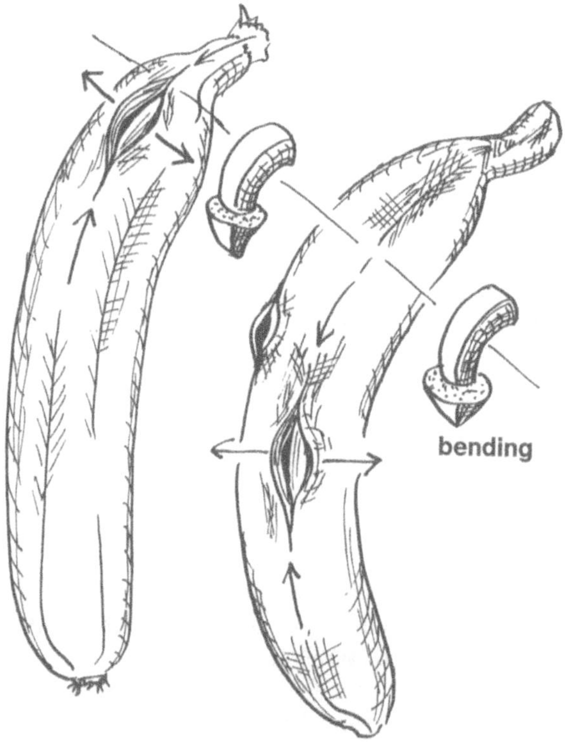

Fig. 65. The convex side of a banana squashes and "cracks" axially if one tries to bend it straight

Since there are neither fibers nor rays in the tangential (tree ring) direction, the risk of splitting cannot be reduced unless by improvement of the tangential glue in the middle lamellae or by tangential-pressure superimposition. While the strength of the glue is restricted by its brittleness, the minimization of the risk of splitting by tangential-pressure superimposition corresponds to the action and effect of the tangential growth stresses (Fig. 64).

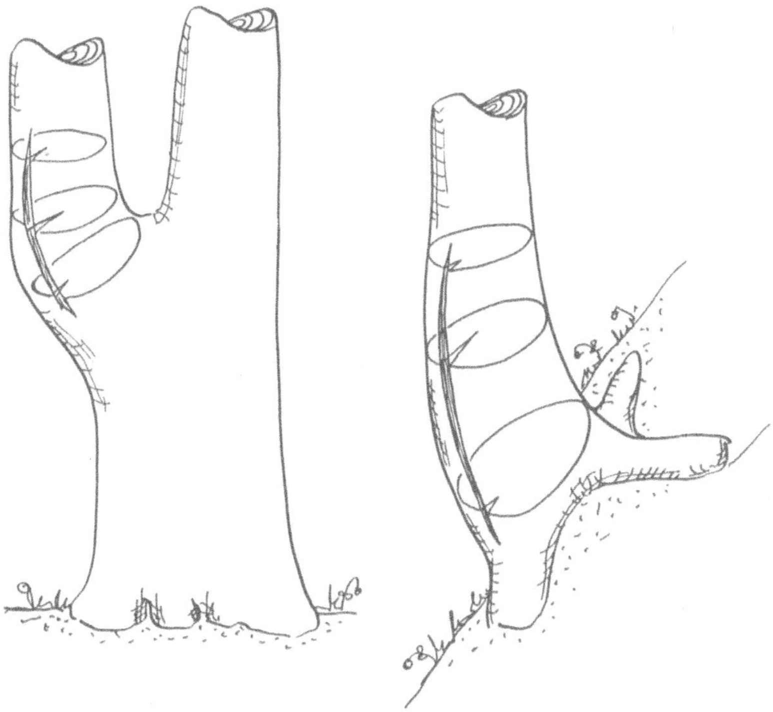

Fig. 66. Bent trees and trees that straighten due to progressive lean tend to develop "banana cracks"

The compressive action of the latter on the tree surface prevents rays from being converted into cracks.

Figure 60 illustrates the behavior of the wood cells under increased internal pressure. The typical longitudinal shortening and transverse swelling of the quasi-axial fibrils accounts for the observed longitudinal tensile stresses and the tangential compressive stresses, held responsible for the case of transverse compressive loading of the rays.

Even in the case of tangential compressive growth stress action, external loading combined with an adverse tree design may effect excessive tangential tensile stresses that convert the rays into cracks. A banana that one tries to straighten by bending (Fig. 65) will break where the bend is most pronounced on its convex side. Slanted trunks or branches that are straightened by external loading may break in the same way (Fig. 66). Such cracks are referred to as "banana cracks" to describe their origin as progressive lean followed by straightening. Characteristic banana cracks in saw cuts of sweet chestnut are shown in Fig. 67.

Searching for answers as to whether the trees detect and fight the risk of cracking, Dietrich and Mattheck (1995) proved that the tangential compres-

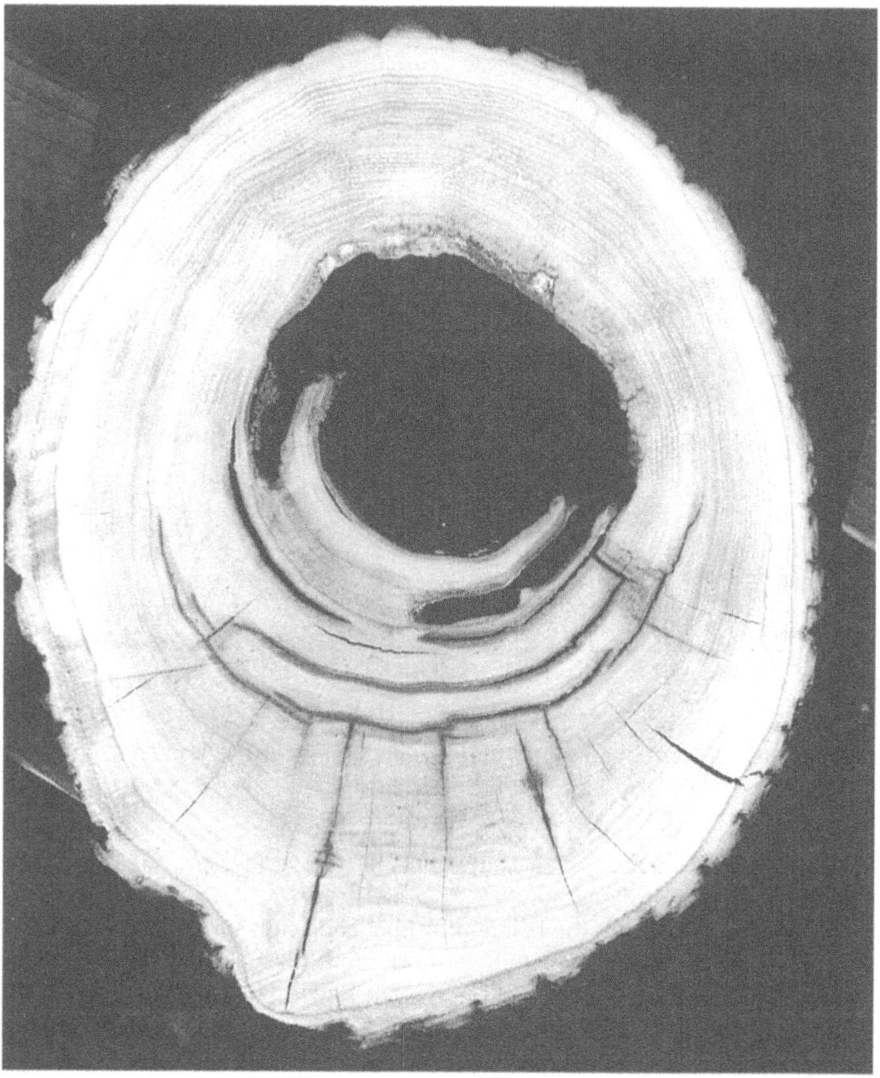

Fig. 67. Cracks due to progressive lean in sweet chestnut

sive growth stresses in angiosperms are greatest where the vertical inclination of the tree is most pronounced, i.e., at the level where tension wood production is most active. Steel pins were punched in the upper and lower sides of saber-shaped beeches found on slopes. Radial saw cuts were made on both sides between the pins. The difference between the spacings of the pins before and after the cut is a measure of the tangential compressive growth stresses (Fig. 68). This seemingly rigorous method will be compared with other methods later.

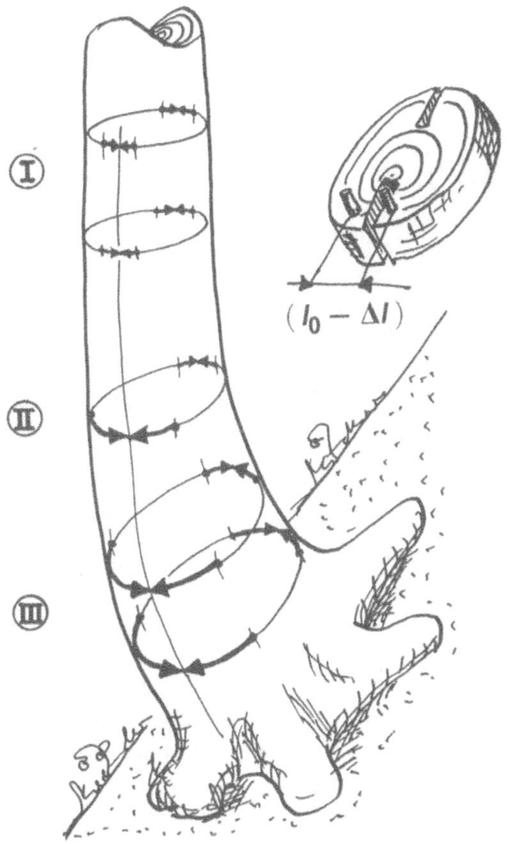

Fig. 68. According to measurements, the tangential growth stress compression of a vertical chain saw cut on the lower side of a saber tree is more pronounced than that of a vertical chain saw cut on the upper side

Some results are shown in Figure 69. Evidently, the tangential compressive stress is greatest where the curvature of the stem is largest, i.e., where the danger of "banana cracks" is highest. Once the tree is upright again, the saw cut is compressed evenly on both sides. From this, one infers that there must be an optimum distribution of the tangential growth stresses, compressing the rays most where the risk of longitudinal cleavage is greatest. The compressive growth stresses in the tangential direction are largest where the external loading and curvatures of the tree or the branches cause the largest tangential tensile stresses. In other words, the largest compressive growth stresses are found where external loading causes the largest tangential tensile stresses. Weak points are avoided where the risk of longitudinal cleavage is distributed evenly.

Because these particular stresses play such an important part in preventing longitudinal cleavage, disastrous consequences are bound to follow when they are neutralized by excessive external loading, or when they subside under constant external loading.

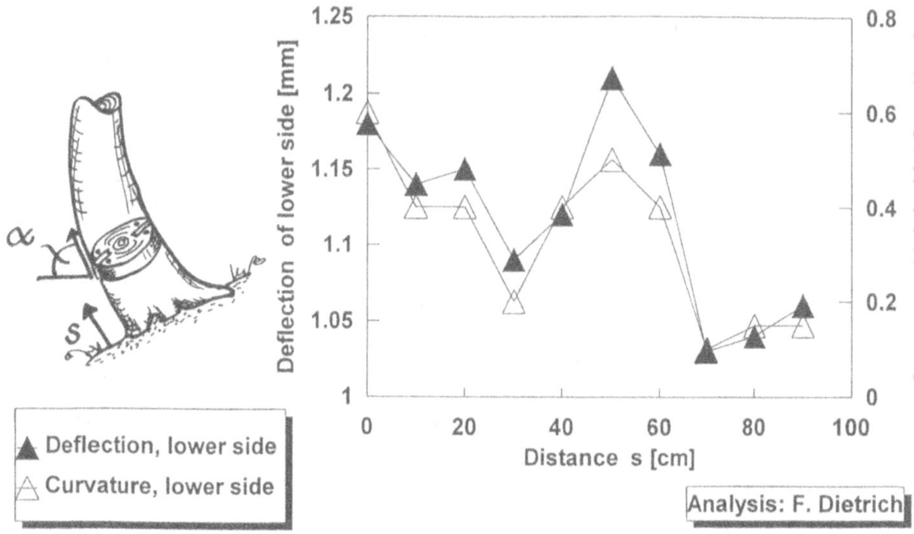

Fig. 69. Tangential displacements after introduction of a saw cut as a function of the tree's curvature

Cracks due to progressive lean (Figs. 66, 67) indicate tangential growth stress neutralization and are signs of excessive external loading or decreasing vigor. The summer drop of branches, however, is a symptom of growth stress subsidence and is due to constant external loading.

Figure 70 illustrates the summer drop of branches by showing the stages of crack formation due to water losses. Deciduous trees lose much water during long spells of hot weather. This loss of water causes a drop in the swelling pressure (internal pressure of the cells), which actually induces axial tensile stresses in the reaction wood on the upper sides of branches. In addition, this swelling pressure is responsible for the tangential compressive stresses that compress the rays (Figs. 60, 64).

When the axial tensile stresses on the upper side of the branch subside due to losses of water, the "tensioning rope" (Fig. 58B), which hinders the branch from dropping, is inactivated. The rope slackens, the branch leans progressively, and the lower side of the branch experiences a greater longitudinal compression, which is accompanied by transverse tensile stresses. Since, in addition, loss of water stops the development of the tangential compressive growth stresses, the rays on the lower side of the dropping branch are drawn apart instead of being compressed tangentially. Obviously, any ray may develop into a crack. The branch breaks or "summerdrops" when such cracks proliferate.

Brittle heart is another phenomenon that can be explained at least partially, by the fact that the rays are dormant cracks. One speaks of brittle heart in the case of brittle trunk cores that show no signs of fungal decay. Brittle

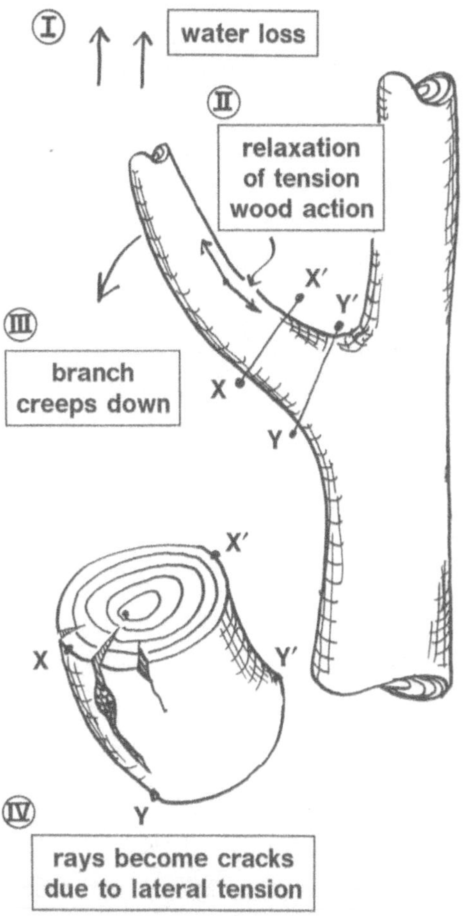

Fig. 70. The summer drop of branches as a special case of crack formation due to water losses. Schematic sketch of a failure

heart develops under the influence of the growth stresses, and is due to the sensitivity of the rays to transverse tension.

Notwithstanding the fact that wood is much more susceptible to compressive than to tensile loading, the propitious axial tensile stresses on the surface of the tree trigger axial compressive stresses in the core on account of the mechanical equilibrium of forces. High values are found (Fig. 62) because these stresses concentrate on a small part of the cross-sectional area of the trunk. The central wood fibers that are subjected to these excessive axial compressive stresses are prone to buckling. In addition to the axial compressive stresses, there are tangential tensile stresses, activating the dormant crack properties of the rays (Fig. 71). The radial tensile stresses, which one finds in the core, act transversely to the tree ring boundaries or upon the weak tree rings of the early wood of ring-porous species.

The core of the tree obviously "pays" for the advantages that its surface derives from the growth stress. Microcracks (Fig. 71B) are bound to develop

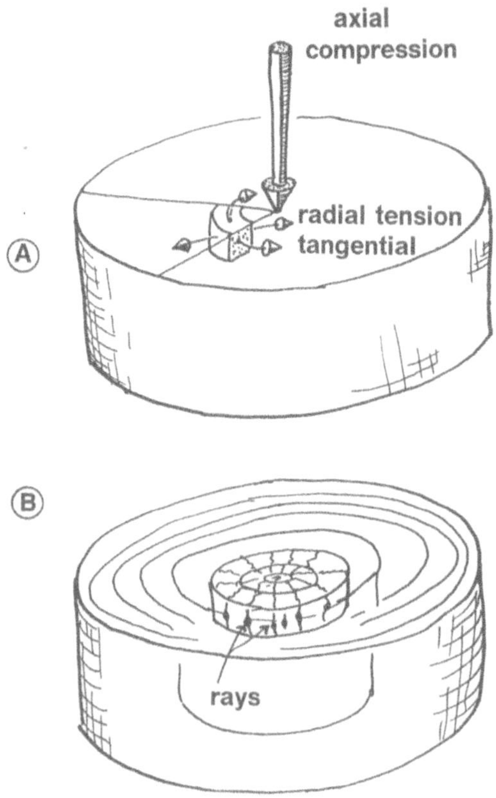

Fig. 71A,B. Brittle heart is caused by radial and tangential residual tensile stresses and by axial residual compressive stresses. **A** Growth stresses in the axial, radial, and tangential (circumferential) directions. **B** Prospective crack lines

due to the detrimental combined action of the axial, tangential, and radial stresses. The macroscopic properties of the wood deteriorate, and the development of brittle heart cannot be stopped once the microcracks accumulate and are joined by other minor or major defects.

The sacrifice of the core over the surface of the stem is easily understood, considering that the bending stresses induced by wind load are smallest in the central areas and even zero at the innermost point of the core. These quasi-no load areas, these lazy parts of the tree, can be neglected or sacrificed for the benefit of busy, active parts. As is shown later, 70% of the radius may be hollow before there is a serious risk of stem failure due to decay. Trees can survive in spite of their sacrifice of heartwood quality. Their species preservation does not depend at all on the condition of heartwood, and they are totally indifferent to man's annoyance at this waste of timber.

The propensity of rays to crack under tangential tension shows especially in the end-splits that may appear when the tree is felled. The end-splits almost always follow the path of the ray.

End-splitting after felling is in the same category of defects or failure mechanisms that men regret but trees regard as irrelevant to their survival.

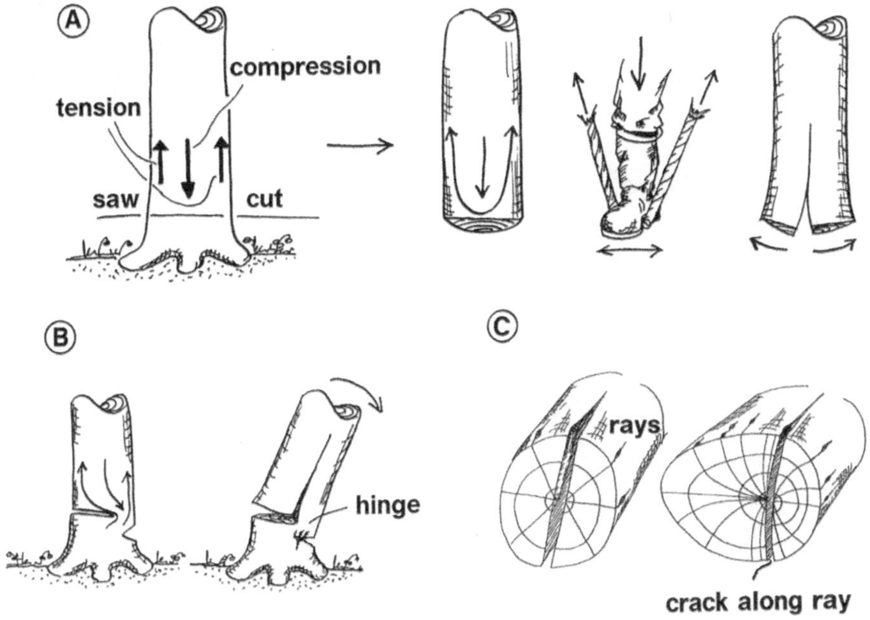

Fig. 72A–C. Two causes of end-splitting afer felling, and end-splitting along the rays. **A** Transverse tensile stresses caused by diverted growth stresses. **B** Transverse tensile stresses induced by the hinge as the tree tips over. **C** End-splitting always occurs along the rays

Trees that end-split after felling are valueless because their trunks exhibit axial cracks up to several meters in length. Beeches and eucalyptus are known to have a particular predisposition to end-splitting. Figure 72 shows the principle of crack formation. In the upright unfelled tree, axial compressive stresses and radial and tangential tensile stresses (brittle heart) act in the core, while tensile stresses are found on the surface.

The new end face of the felled tree disturbs its mechanical equilibrium. The axial tensile stresses must be diverted transversely to the axis of the trunk along the sawn end face. Finite-element calculations were made to prove this reorientation (Mattheck and Walther 1991). One may imagine the compression-loaded tree core as "stepping" into a sling that supports its diverted tension. The additional transverse tensile stresses produced as the axial tensile stresses are diverted may start axial cracks running up the stem (Fig. 72A). Tree-felling under real conditions reveals some complex processes in addition to those found in the theoretical model. Notch stresses occur because the axial force flow must be diverted transversely around both the notch and the saw cut through the narrow hinge. Moreover, the bending stresses to which the hinge is exposed as the tree tips over may induce failure because they are diverted across the grain (Fig. 72B).

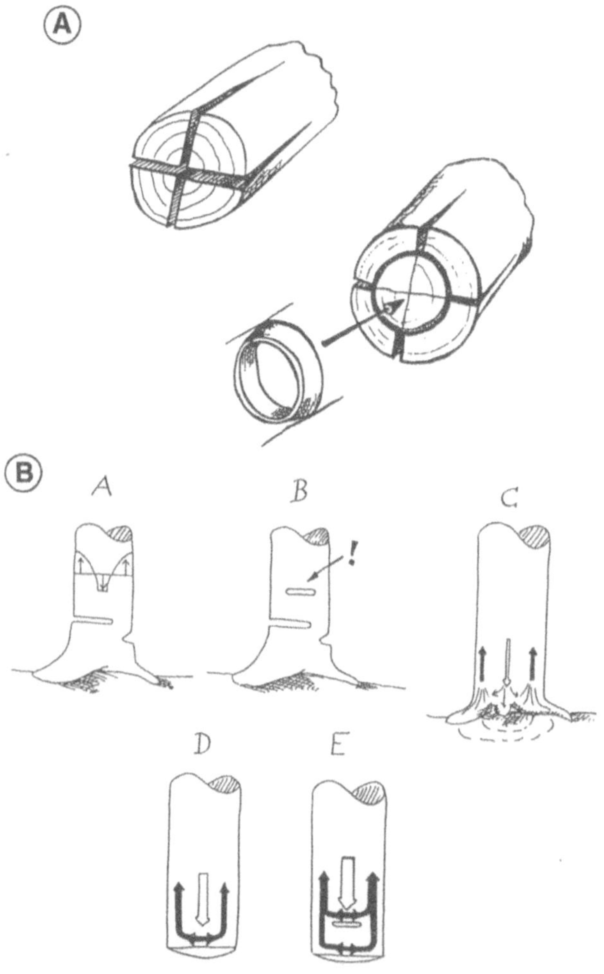

Fig. 73A,B. Two moderately successful methods to avoid end-splitting or limit crack formation. (Mattheck 1991). **A** Conical compression ring. **B** Additional keyhole cut

In spite of all these disadvantages, the design of each tree is as perfect as can be. Trees strive for their optimum survival; they do not arrange to die through optimum felling.

Nevertheless, it is more than regrettable when the trunk of a 160-year-old beech, cared for by generations of solicitous foresters, is devalued by end-splitting. In search of suitable countermeasures, experts in Europe have been using so-called S-hooks to limit the propagation of existing cracks. S-hooks have been applied for decades, but have proven to be rather inefficient because they act in only one direction. The problem of crack propagation in another direction was solved by a conical compression ring (Mattheck 1991), which is punched into the cracked trunk (Fig. 73A). This ring effectively limits crack propagation and even recompresses existing cracks. These compression rings are, however, difficult both to drive in and to remove. They

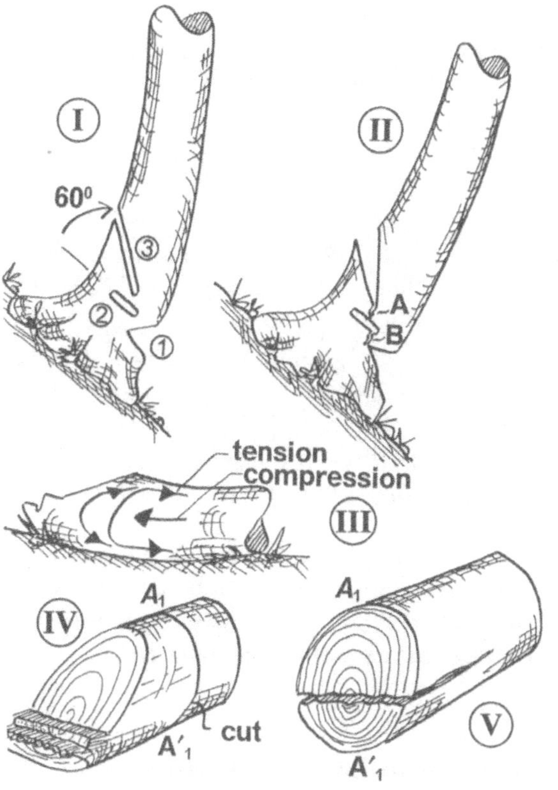

Fig. 74. The best end-split avoidance method found so far during field tests. (Mattheck 1991)

are relatively expensive into the bargain, and are applied only *after* crack formation.

In the search instead for a preventive measure, a felling technique was developed that diverts the force flow at two levels. This two-level diversion is achieved by introducing an additional keyhole cut (Fig. 73B). This technique has been successful on the whole. It was dropped, nevertheless, because cracks flashed over sometimes into the stress-relieving cut, propagating themselves and ruining the stem completely.

Figure 74 shows the principle of the best method developed so far to avoid end-splitting. The preventive effect of slanted cuts has been known for some time (Tantichaiboriboon and Cook 1976). Looking for examples in nature, one finds that the conical ends of the trunks seen in beaver's lodges are mostly uncracked. FEM calculations (Mattheck and Walther 1991) also prove that slanted cuts considerably reduce the transverse tensile stresses acting on the saw cut.

Slanted cuts definitely limit the risk of crack formation due to growth stresses. They do not suffice, however, to eliminate the cracking induced by the lever action of the hinge as the tree tips over (Fig. 72B). Experiments carried out with beeches show that further steps are required to control cracking. Three cuts are recommended: the felling notch, a keyhole cut that minimizes the hazardous lever action of the hinge, and a slanted cut (Fig. 74). When the slanted cut has reached a certain depth, a crack flashes over from the tip along the grain into the keyhole cut. The remaining narrow hinge does not introduce major splitting forces into the falling stem. The felling notch should not be too acute, otherwise crack-inducing shear stresses may develop when the faces are crushed as the tree tips over. In addition, the hinge should be wide enough to give a controlled felling that does not endanger the loggers. This three-step technique was tried on numerous beeches. Most of these trees were growing on slopes and were particularly prone to splitting due to pronounced tension wood development. End-splitting was avoided throughout. A more or less indirect proof of the method's success was produced by cross-cutting the conical ends of some trees right after felling. The fact that end-splitting set in immediately in 80% of the trunks thus treated reveals that trees with conical ends must be seasoned until all growth stresses have subsided.

There are certain drawbacks. For example, thick trees need keyhole cuts coming from two sides, the conical ends of the trunk must be removed by sawing, and the stub should be leveled after felling. In spite of this time-consuming extra work, the advantages of the method recommend it.

8.4 Selected Techniques for Measuring Growth Stresses

8.4.1 Visual Evaluation

Purely visual inspection is suitable only for evaluating growth stresses in the reaction wood of leaning trees or branches whose tension or compression wood produces unsymmetrical growth stresses, for directional corrections. The observer can see whether the leaning stem or branch still resists the force of gravity through negative gravitropic growth, i.e., whether there is a tendency to straighten or to lean (Fig. 75). The upper side of the branch shown in Fig. 75A exhibits folds like those in the bend of the human elbow, or in the sleeve of a leather jacket. These folds indicate an increasing upward movement, compressing the bark on the upper side of the branch and stretching it on the lower side to overcome the force of gravity. Branches with a tendency to lean (Fig. 75B) show folds and compressed bark on their lower sides. On the lower sides of thick-barked trees, one detects bellows-type bark formations, comparable to the folds under the arm of a jacket. The bark on the tension-loaded upper sides of the branches of thick-barked trees (Fig. 75C) may even peel off. It is mostly thinner and looser than that on the compression-loaded

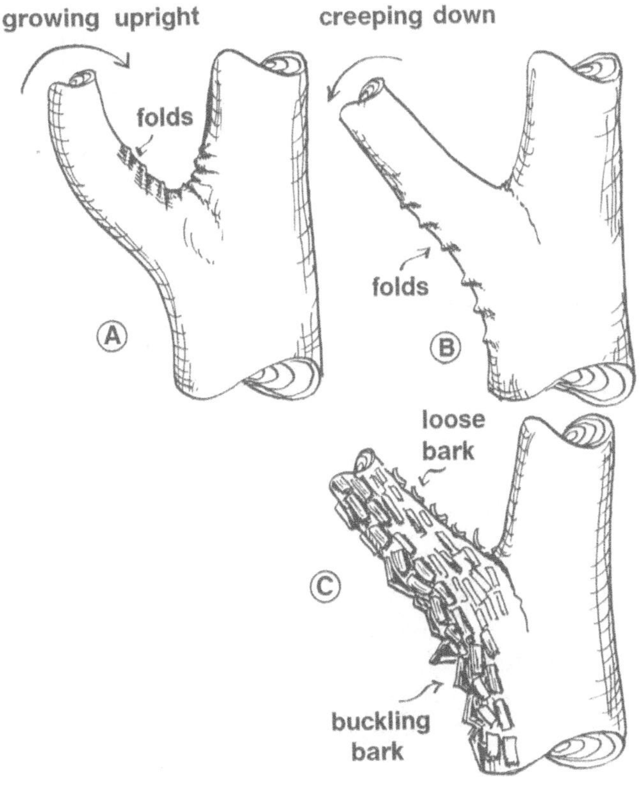

Fig. 75A–C. Visual evaluation of tension wood in branches. **A** High growth stresses pull the branch into position. **B** The branch progressively leans. **C** Thick-barked trees exhibit bellows-type folds in the bark on the compression side and loose bark and bare spots on the tension side of bending

lower sides, and it lifts off very easily. Thick-barked branches with pronounced lean are often stripped and bare on their upper sides.

Growth stresses in leaning stems are evaluated in the same way as those in leaning branches (Figs. 76, 106). Extremely leaning trunks are often found with slip lines in their centers. The two halves of such trees may slip on each other like the pages of a book when it is bent. Their radial rays interlock the two halves against shear load except in the horizontal plane of the slip line. For this reason, shear cracks are always found at right angles to the main wind direction. Thick-barked leaning trees, like thick-barked leaning branches, show bellows-type folds on their lower sides and loose bark on their upper sides.

Weak points in metal components are detected by applying crackle varnish before subjecting the specimens to loads in a testing machine. The coat

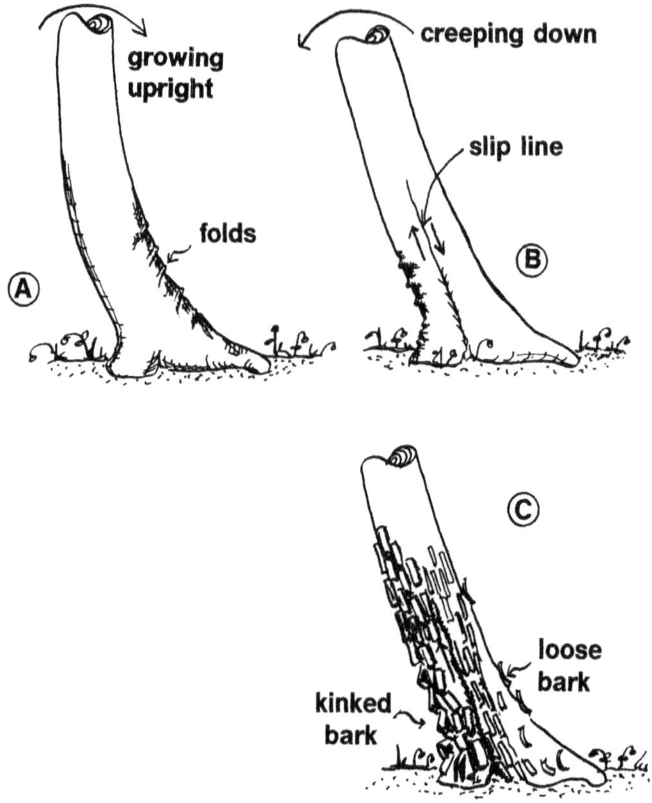

Fig. 76A–C. Visual evaluation of axial tension wood in slanted trees. **A** High growth stresses pull the stem into position. **B** The slanted tree progressively leans. **C** Rough bark shows more pronounced symptoms

lifts off first where the load is greatest. In that sense, the bark is the crackle varnish of the tree, and transverse cracks, stripped parts, and bellows-type folds should be looked for if one wants to evaluate the safety of a tree. Such areas can be assumed to be either thin-walled or exposed to highest stresses. Bark inspection is a helpful, but rough and relative means of visual evaluation.

8.4.2 Measurement of Growth Stresses with an Increment Borer

The axial tensile and tangential compressive stresses acting on the surface of the tree lead one to assume that borer cores flatten in the tangential direction. The core cross section, which is compressed laterally by the growth stresses before removal, is expected to expand tangentially (Fig. 77).

These seemingly plausible assumptions do not, however, actually, occur, and the deformation is just the reverse of what is expected (Polge and Thiercelin 1979, Archer 1987). Figure 78 explains why the circular edge of the increment borer is bound to deform during drilling. The greatest forces come

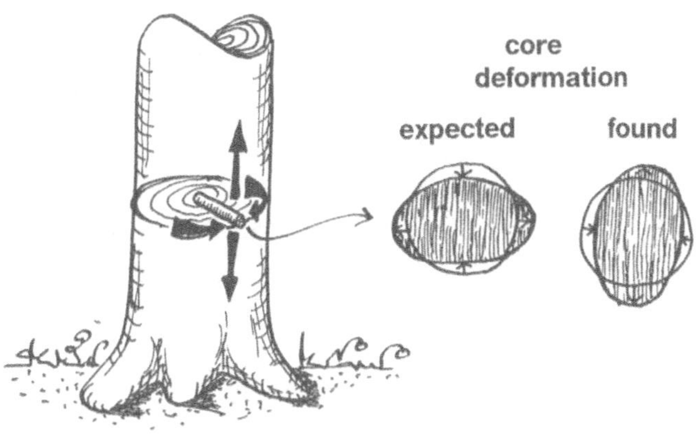

Fig. 77. Wasted efforts at determining growth stresses from increment borer cores

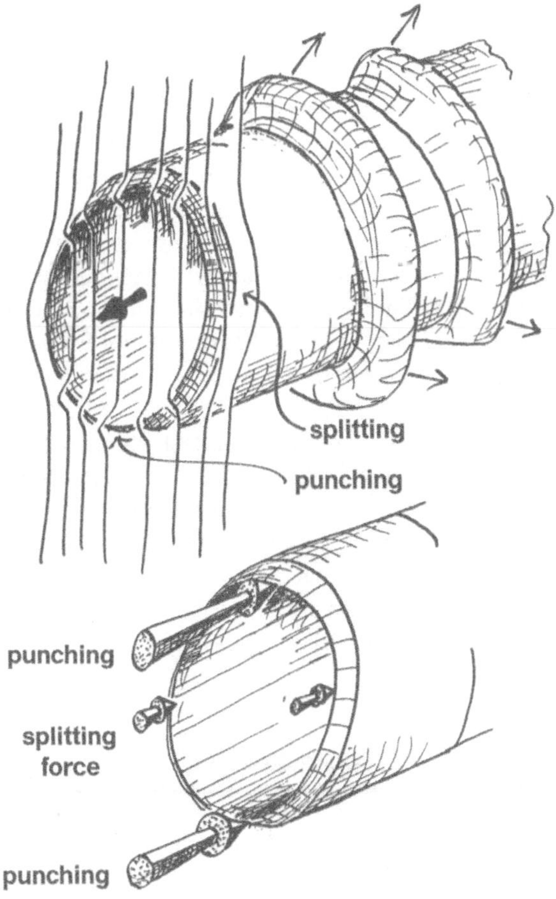

Fig. 78. Inhomogeneous loading of the edge of the increment borer; characteristics and origin

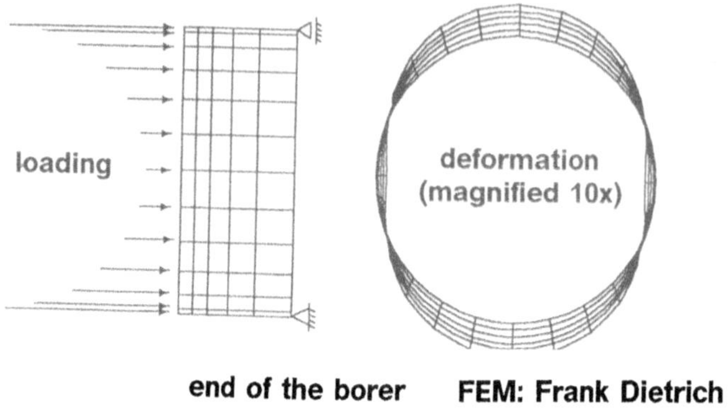

Fig. 79. Numerical finite-element verification of the ovalization of borer cross sections as shown in a deformation plot

from the upper and lower sides of the edge. They are required to penetrate the wood fibers transversely. Between the fibers, the edge passes smoothly and easily overcomes the negligible cleaving resistances. These differences in the required penetration forces mean that the edge of the increment borer is not loaded uniformly (Fig. 78). The resulting deformations, calculated by Frank Dietrich (Kernforschungszentrum, Karlsruhe) by means of the finite-element method, are plotted in Fig. 79. The borer edge develops into the expected vertical oval. The degree of ovalization depends on certain boundary conditions; for example, a blunt borer experiences greater penetration forces and ovalizes more than a sharp borer, a greater cleaving resistance reduces core ovalization, and differences between the two lateral borer edges may occur if one edge cleaves a ray longitudinally while the other passes between the fibers. In a nutshell, borer edge ovalization depends on the borer quality, on the tree species, on individual tree characteristics, and on the point chosen for drilling.

The ovalization of the edge, no matter how pronounced, combines with the opposite ovalizations induced by growth stresses. This was found to result in ovals characterized by longer vertical semiaxes (Fig. 80). Core drying even increases this typical axial relation because drying wood shrinks more in the tangential direction. It is not recommended to make measurements some time after core removal. Such measurements would be imprecise, due to superimposed borer deformation, growth stress, and drying stress effects.

8.4.3 Growth Stress Measurement with Prismatic Samples

Prismatic samples are obtained by inserting two staggered steel pins and by removing the wood surrounding them. The pin spacing is measured before and

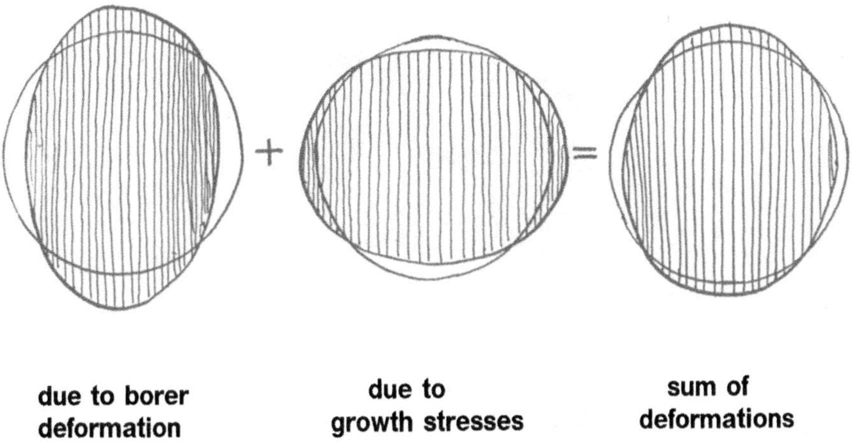

Fig. 80. Superimposition of counteracting borer deformation and growth stress ovalizations

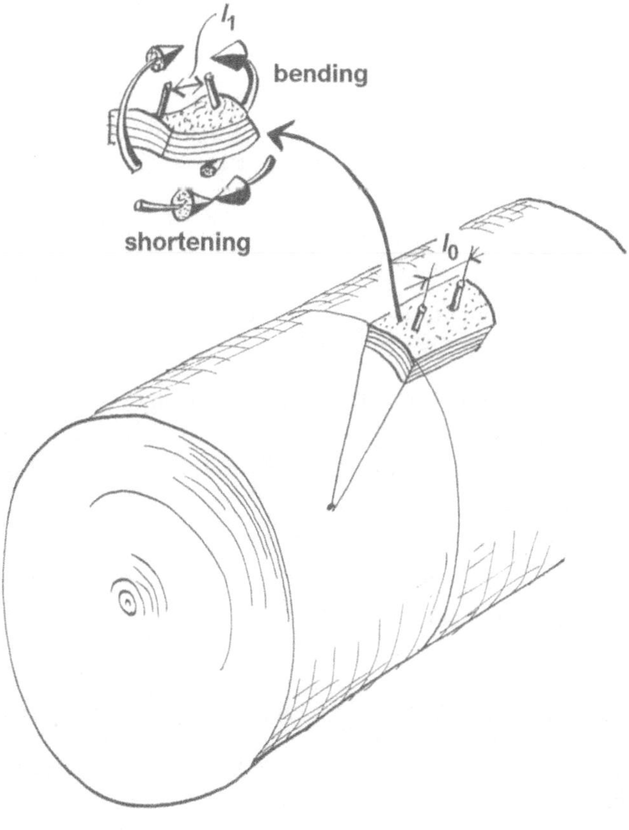

Fig. 81. Tensotast measurement of the longitudinal deformations that develop as one removes a prismatic element. (Archer 1987)

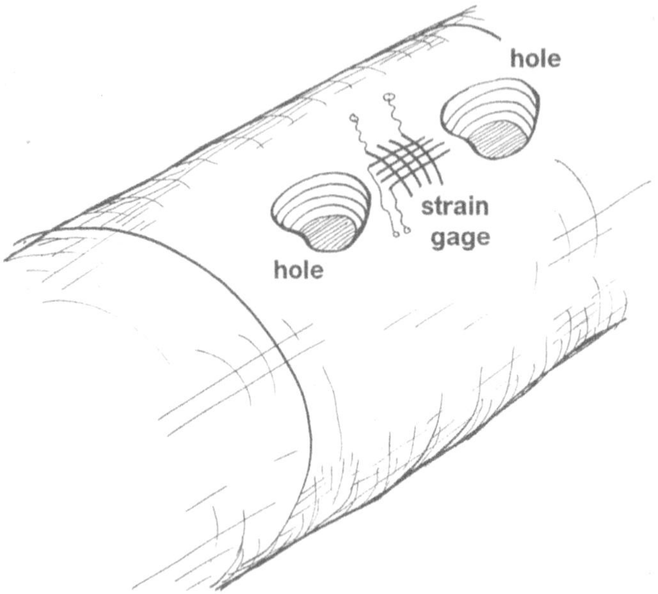

Fig. 82. Strain gage measurement of the deformations that develop in the presence of two vertically offset bores

after sample removal for calculation of the growth stresses (Nicholson 1971, Atherton 1978). Fairly good results are obtained, especially if one considers the radial growth stress gradients that are revealed by additional sample bending (Fig. 81).

8.4.4 Growth Stress Measurement Between Two Pocket Holes

Pocket holes are drilled above and below crossing strain gauges fixed to the tree (Archer 1987). Undesirable bending stresses (Fig. 82) can be suppressed relatively easily, but the lateral supporting action of the remaining wood will slightly reduce the measurements.

8.4.5 Single-Hole Growth Stress Measurement

Several steel pins are arranged in a circle, with one pocket hole drilled in the middle (Fig. 83). The different movements of the pins relative to one another allow one to calculate the respective strains. Archer (1987) gives very detailed descriptions of this method and of other techniques.

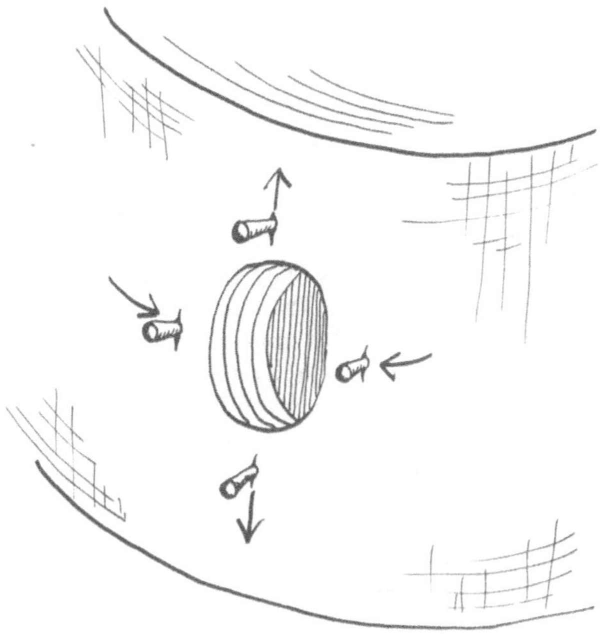

Fig. 83. Measurement of the deformations that develop in the presence of a single bore

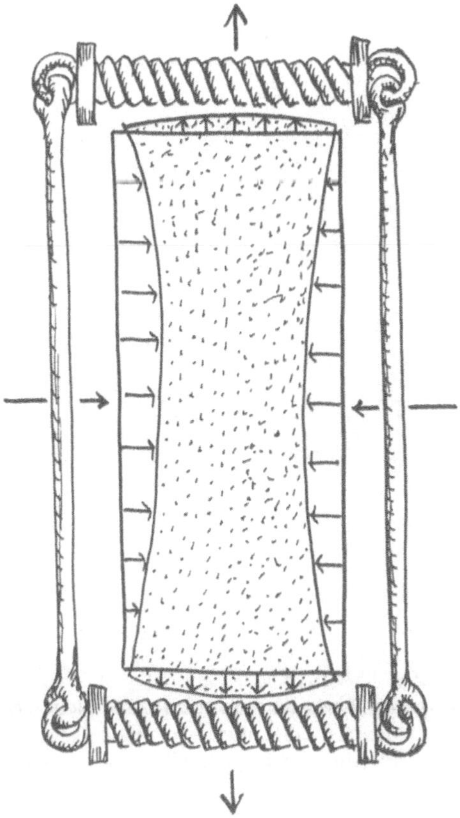

Fig. 84. The informational value of the real growth stresses is qualified by the lateral supporting effect and by the notch stresses

The methods described above give information about the distribution of growth stresses in trees. However, the value of this information is qualified by certain disturbing factors.

The lateral supporting action of the wood that remains, when one takes a prismatic sample, leads to undesirable gradients of deformation (Fig. 84). The tensile bars represent the remaining wood that impedes the elongation of the vertical lateral edges of the hole. The horizontal threaded rods represent compression bars that prevent the upper and lower edges of the rectangular hole from being compressed. The tensile bars and compression bars represent the surrounding material's "neighborly help". Their supporting effect is irrelevant to sample measurement (Fig. 81), but impairs the precision of results when, for example, pocket holes are measured directly. The precision of the results is also marred by local orthotropic variations in the Young's moduli. One obtains varying deformations because notch stresses change depending on the directional stiffnesses. The comparative results presented by Archer (1987) are quite convincing in spite of these deficiencies. The results may not be precise, but can be assumed to be useful and basic approximate values.

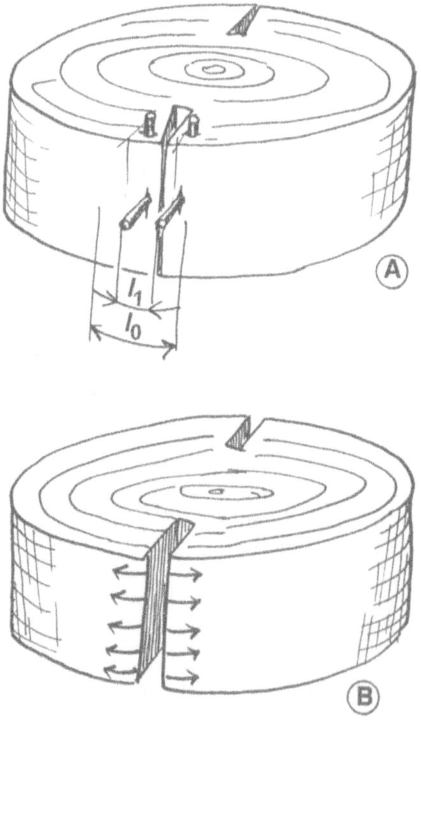

Fig. 85A,B. Radial saw cuts in slices of green wood (**A**) are a rigorous means of evaluating tangential growth stresses. After drying, the saw cut will open up (**B**)

8.4.6 Tangential Growth Stress Measurement on Radial Saw Cuts

Neglecting the longitudinal values, Frank Dietrich (Kernforschungszentrum, Karlsruhe) used radial saw cuts to measure tangential growth stresses. The tree is felled and is cut into slices (Fig. 85). Two steel pins are then hammered into the cut surface near the bark, and a radial saw cut is made between the pins. The difference between the pin spacings before and after radial sawing is a measure of the tangential compressive stresses. It seems that the precision of results is impaired because slicing across the longitudinal growth stresses also largely eliminates the relevant transverse stresses (Poisson's ratio). However, these transverse stresses were found to be insignificant. If transverse stresses were decisive, the measured end face and circumferential values would differ widely. Since comparable results were obtained for slices of different thicknesses, the tangential growth stress values can be assumed to be precise enough. The tangential compressive stresses in the middle levels of slices are slightly higher than those at the ends. These differences are quasi-negligible and do not falsify the results either.

The results obtained applying the radial saw cut method (Fig. 68) are very plausible. This method allows one to measure major deformations that are otherwise hidden or reduced due to the lateral supporting effect of the wood remaining when samples are taken. The radial saw cut method may be somewhat rigorous. It is, nevertheless, preferred and recommended because of its advantages over other techniques.

Summary

- Growth stress development is triggered by a pneumatic principle.
- Growth stresses counteract and reduce the hazardous stresses in trees.
- Axial tensile stresses prevent fiber buckling under compressive loading, while tangential compressive stresses counteract the cracking properties of the rays.

9 Stress-Controlled Strength Distribution

9.1 The Wood-Concrete Analogy

Troll (1959) and other authors discuss the resemblance between wood and reinforced concrete. Figure 86 shows the underlying principle. Whereas reinforcing steel and cellulose are characterized by high tensile strength, concrete and lignin resist compression but cannot bear much tension.

Figure 87 represents a modified version of this model (Mattheck 1995). Figure 87A shows the section of a ring-porous tree ring with large earlywood vessels, and the exploded view of a wood cell. The primary cell wall and the middle lamella form a lignin chimney (Fig. 87B) which houses the cellulose vein of the secondary wall. The rays are shown as I-beams. The brittle but pressure-resistant lignin chimneys are particularly thick where the tree is exposed to compressive stresses, e.g., on the lee side of the preferential wind direction or on the lower sides of leaning trees. The lignin located in the secondary cell wall is imagined also to occur in the lignin chimney; a separation is necessary to obtain this simple model.

The reaction wood (compression wood) of conifers is a typical example of thick chimney walls rich in lignin that serves to prevent failure caused by buckling. The tension wood of deciduous trees is not as rich in lignin, but is characterized by thick cellulose veins, forming the gelatinous layer. Tension wood easily yields under bending stresses but is ideally adapted to tension.

The model facilitates access to very extreme decay mechanisms. The delignifying selective white rot which (shown idealized in Fig. 88A) decomposes the lignin chimneys but does not affect the soft cellulose veins, i.e., the tensile strength of the wood. Selective white rot also destroys the lignin matrices of the I-beam or ray cross sections, leaving them with nothing but cellulose veins. The rays do not lose their tensile strength but become more sensitive to bending. Pieces of wood degraded by white rot are soft and kneadable. They consist, so to speak, of an easily bent, nylon-type material that resists tearing.

Brown rot has just the opposite effect (Fig. 88B). While the cellulose tubes are decomposed selectively, a stiff but brittle structure of clotted lignin tubes remains. This remaining structure is stiff and resists bending, but breaks under insignificant loading. Trees infested with brown rot may tip over without warning. Failure due to brown rot can seldom be predicted. Simply speaking, brown rot turns the wood into a stiff but brittle biscuit, whereas successive white rot converts it into a soft, but very tough, leathery sandwich. Between

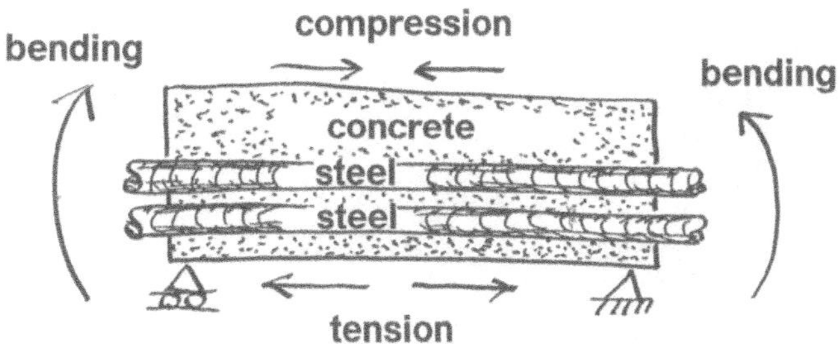

Fig. 86. The wood-concrete analogy

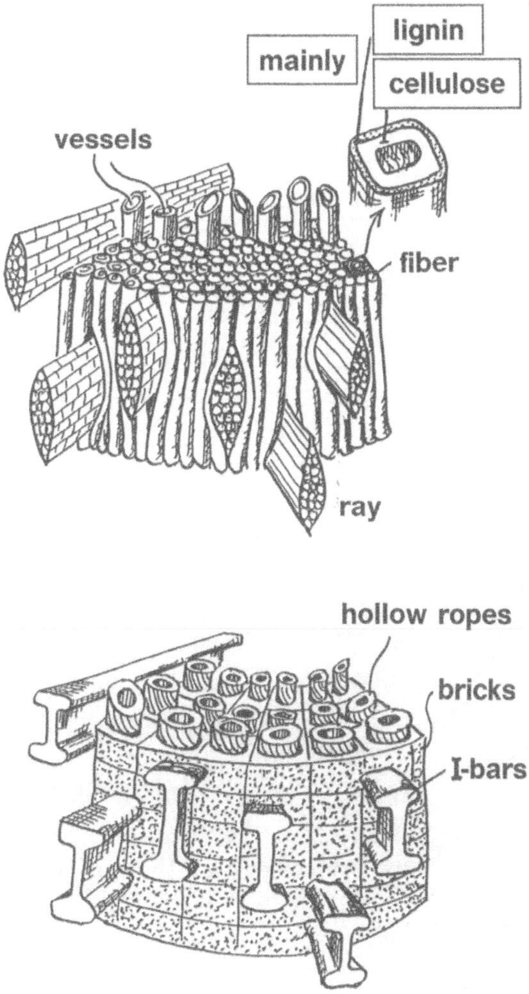

Fig. 87A,B. Wood resembling a longitudinally and transversely reinforced brick wall. **A** Section of a ring-porous tree ring. **B** Mechanical substitute model

these two extreme types of decay lies a continuum of all the other forms of decay, destructive to both lignin and cellulose simultaneously.

Soft rot (Fig. 88C) is a special type of white rot. Schwarze et al. (1993) proved that soft rot may also occur in living trees. The characteristic local

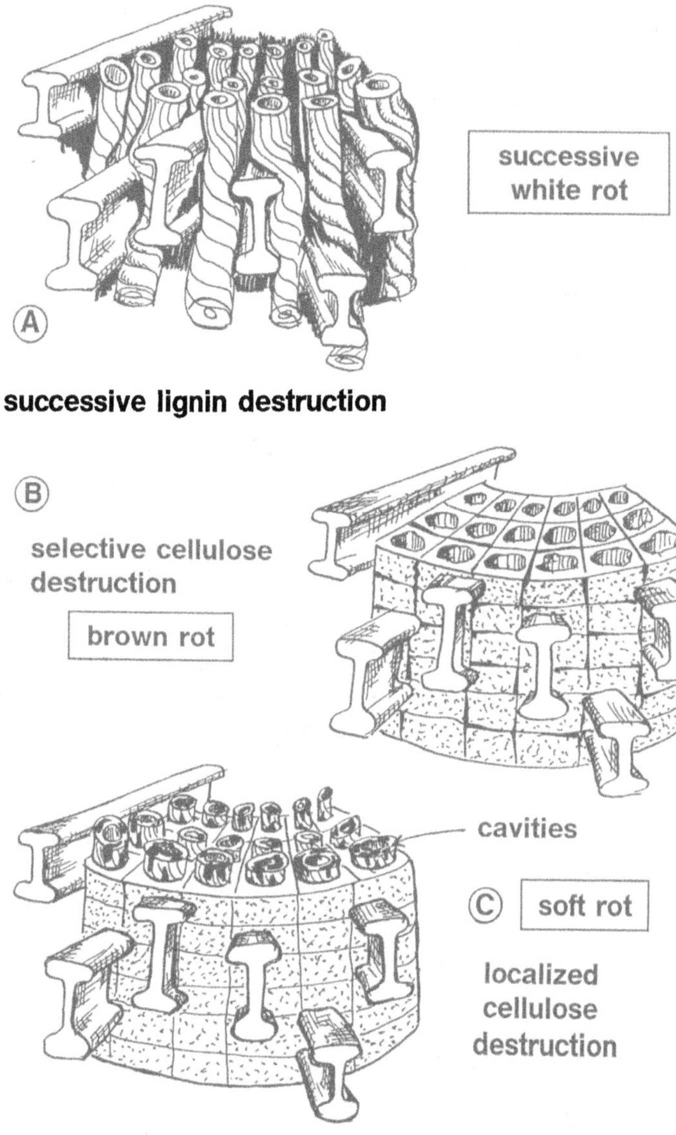

Fig. 88A–C. The three main patterns of wood decomposition caused by fungi. **A** White rot with selective delignification. **B** Brown rot with pure cellulose decomposition. **C** Soft rot causing pitting in the secondary wall

cavities in the secondary cell walls decompose the wood but do not instantaneously destroy the cellulose veins. Soft rot can also, however, cause brittle fracture as a consequence of decay. Failure caused by brown rot or soft rot is often accompanied by ceramic-like fracture surfaces without fiber structures. This is due to the fact that the cellulose veins are decomposed, while the brittle, hollow lignin chimneys remain.

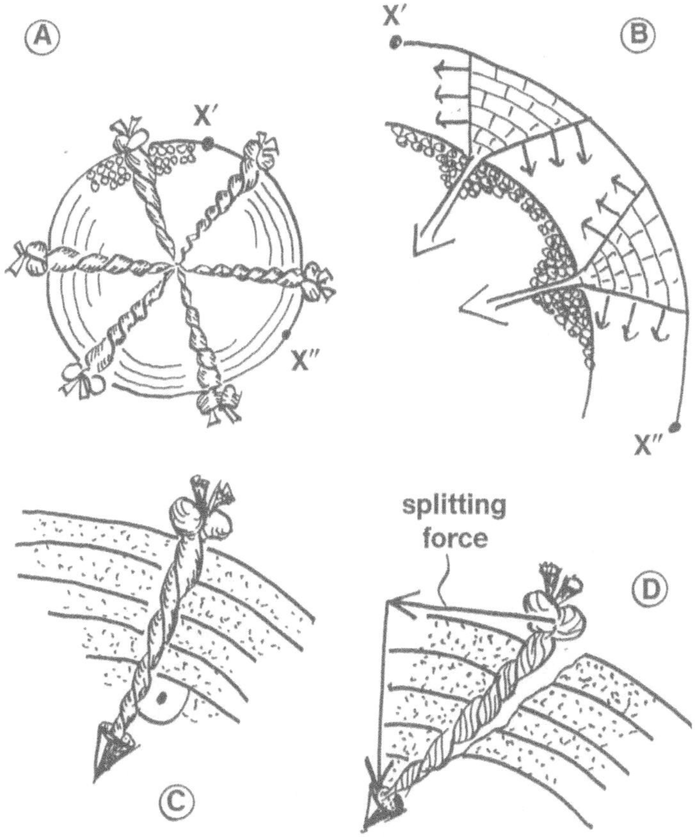

Fig. 89A–D. Rays as radial bracings. The advantages of rectangular ray-and-tree-ring crossing: splitting forces are minimized. **A** Rays as radial bracings; **B** triangular ray ends in lime trees; **C** vertical ray-tree-ring crossing avoids splitting forces; **D** crossing out of perpendicular introduces tangential splitting forces

The rays are of special importance as regards the strength of the wood. They function as radial bracings which delay longitudinal cleavage. Their vertical tree ring crossing, even in the case of exotic, unusual arrangements, minimizes the cleaving forces. Figure 89 proves the advantages of vertical ray-and-tree-ring crossing by showing that crossing out of the perpendicular introduces tangential cleaving forces into the tree ring. According to Troll (1959), lime trees enlarge their rays peripherally in the bark while forming knot-like triangular anchors that compress the wood. In addition to functioning as radial bracings, the rays serve as cross-bolts which interlock adjacent tree rings against axial shear (Fig. 90). There are different interlocking strategies. Thin rays, which are relatively sensitive to bending, experience bending stresses in the area of the earlywood (e.g., ash trees, Fig. 91A). Cross-sectional lignin dyeing shows that the rays of ash trees are well adapted to such

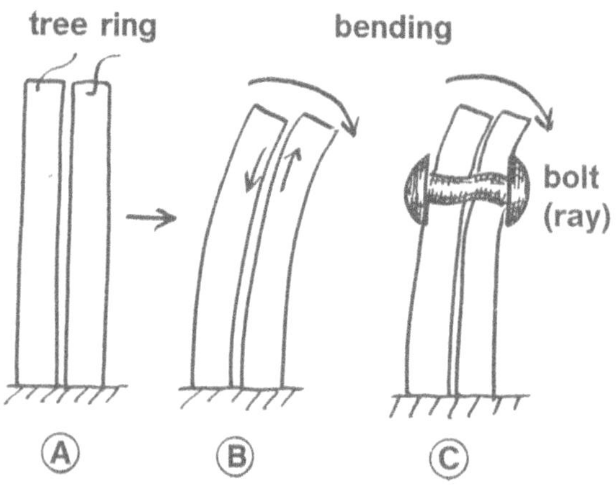

Fig. 90A–C. Rays as transverse interlocks against axial shear between the tree rings

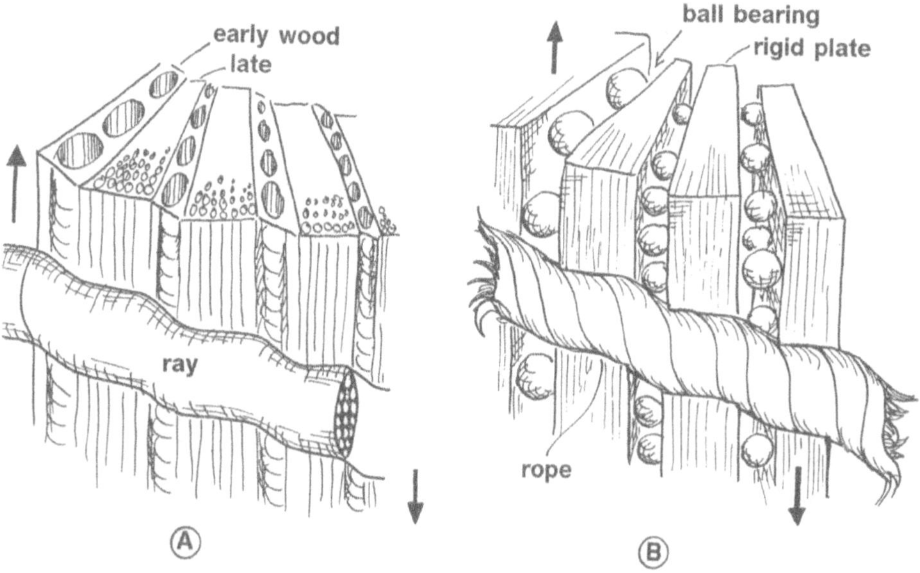

Fig. 91A,B. Flexible interlocking strategies in ring-porous wood (ash trees)

bending (Mattheck and Schwarze 1994). The dye reveals that the ray cross sections resemble I-beams, with stronger lignification in their upper and lower parts.

Figure 91B represents a slightly overdrawn mechanical substitute model. The pores of the earlywood function as ball bearings, which loosely connect the corresponding latewood layers. The rays are elastic ropes, which interlock the bearings. The flexible interlock of this structure reminds one of a wicker chair. It may have been due to this flexible rather than brittle interlock

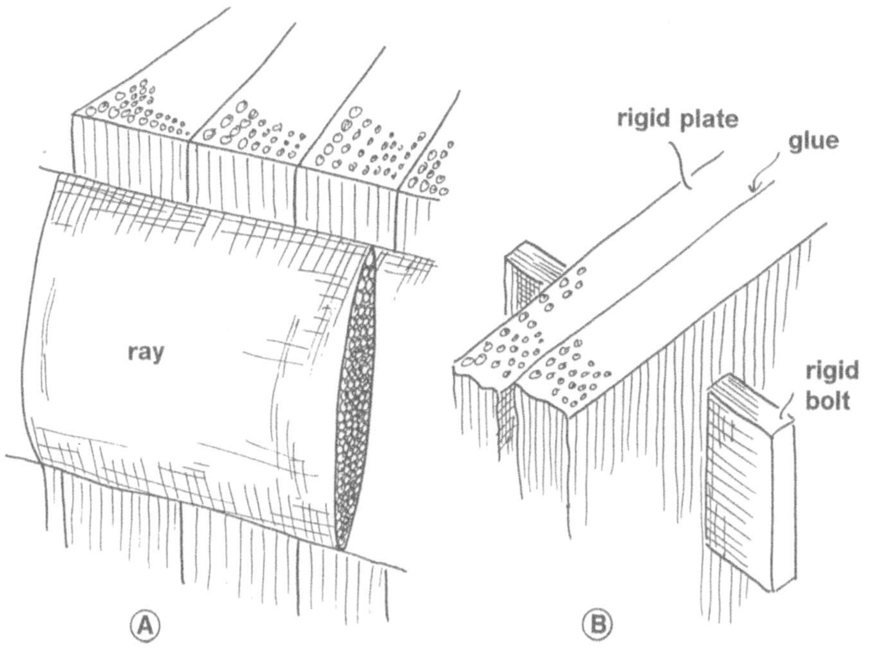

Fig. 92A,B. Rigid interlocking strategies in diffusely porous wood (beech trees)

of the fibers that ash wood was the preferred material for shafts of battleaxes and hunting spears during the Middle Ages.

Altogether different interlocking strategies are found in oak trees and in beech. The rays of oak may be some centimeters in height. They bridge the ring-porous earlywood, making rigid interlocks. The several-millimeters-high rays of beech behave in a similar way. Such trees use "radial boards" to interlock tree rings (Fig. 92), and their wood is brittle compared with that of ash trees.

There are some transitional methods in addition to the flexibility and rigidity strategies. Ring-porous oak, for example, pays for the advantages of its rigidity strategy with exceptionally high ray cross sections, which must rigidly bridge the earlywood gap. Moreover, the tree ring boundary seams that the rays exhibit play a special part (Fig. 93). At the beginning of a new vegetation period, material must be attached where ray growth ceased in autumn. Such transition seams are inferior to continuous rays, just as welded engineering components are inferior to weldfree components. The tree makes up for this weakness by enlarging its ray cross sections in the area of the seam, i.e., at the tree ring boundary. More material of inferior quality is used to reinforce the weak point. There is a very good example in (Shigo 1994). Such seams are also related to the bulges of grafted trees.

The section below discusses the relevance of the transverse strengths of wood to the carrying capacity of trees.

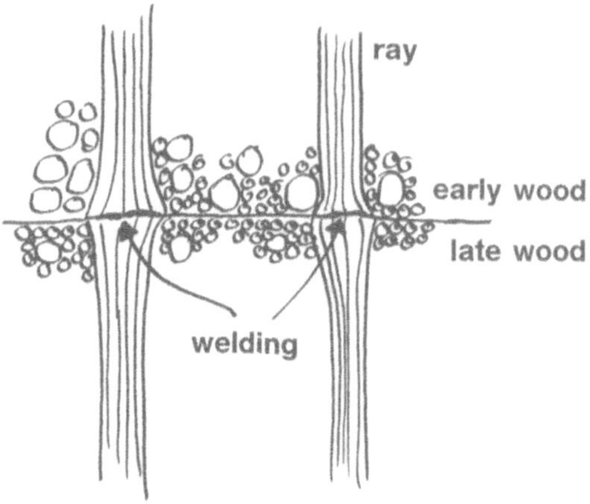

Fig. 93. Enlargement of rays at the tree ring boundary

9.2 Radial Transverse Strengths and Hazard Beams

A pocket-sized wood testing device, the fractometer, was developed (Mattheck and Breloer 1993, Mattheck and Bethge 1993) to break a 5-mm increment borer core under bending loads. Figure 94 shows the basic features and the mode of operation of the fractometer. The fractometer serves not only to evaluate intact trees, but is also particularly suited for examining decayed wood, i.e., for evaluating hazardous trees. The core is inserted, and is loaded by turning the spiral spring. The hammer is pressed down on the specimen and bends it. The bending angle shows in the fractometer window; the load is indicated by a trailing pointer. When the critical bending stress is reached, the sample breaks as fibers separate on the tension side of bending. To ensure natural ray bending, care must be taken to insert the core with the wood fibers in the plane of bending, i.e., running parallel, not vertical, to the face of the fractometer.

The rigid but brittle, biscuit-type cores of trees infested with cellulose-decomposing brown rot will break under insignificant bending at small fractometer angles. Successive white rot delignifies and leaves tough but soft cellulose. Large fractometer bending angles and relatively great fracture stresses indicate the flexible but tough cores of trees whose breakage is predictable and undramatic. The fractometer allows the rough classification of failure potentials from brittle fracture such as in brown rot, and ductile fracture as in successive white rot. This classification helps the expert assess the safety of urban trees. The weakest spot in the trunk has to be identified by the arborist's experience, and there the core must be removed.

Figure 95 represents some magnified cross sections of rays, shown in identical scale for comparison. Some typical fractometer values are shown as fractometer units (FU), i.e., 100 FU ≈ 26 MPa = 26 N/mm^2. Evidently, higher fractometer values are obtained for hardwood, which has several rows of

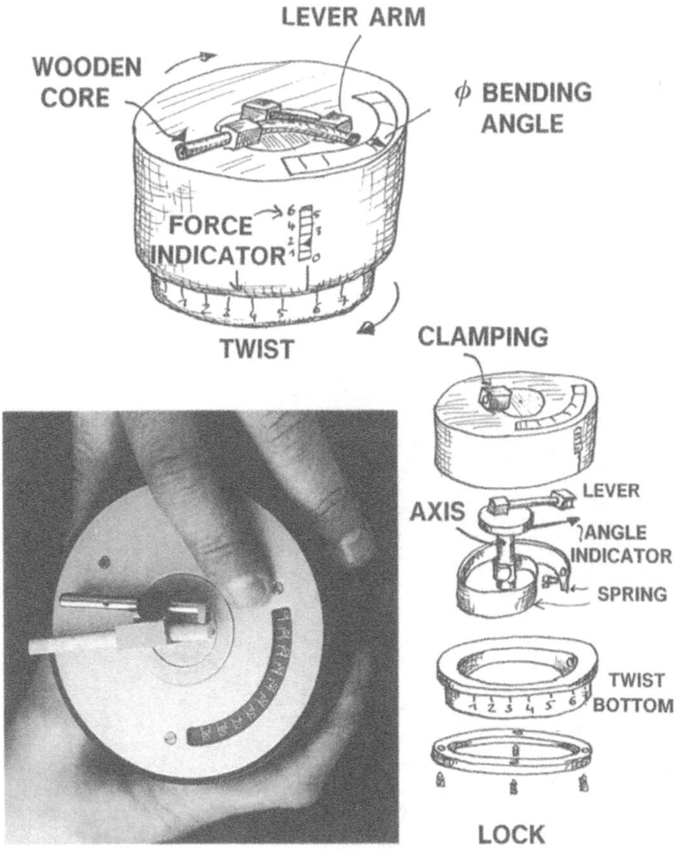

Fig. 94. Fractometer measurement of the transverse bending strength

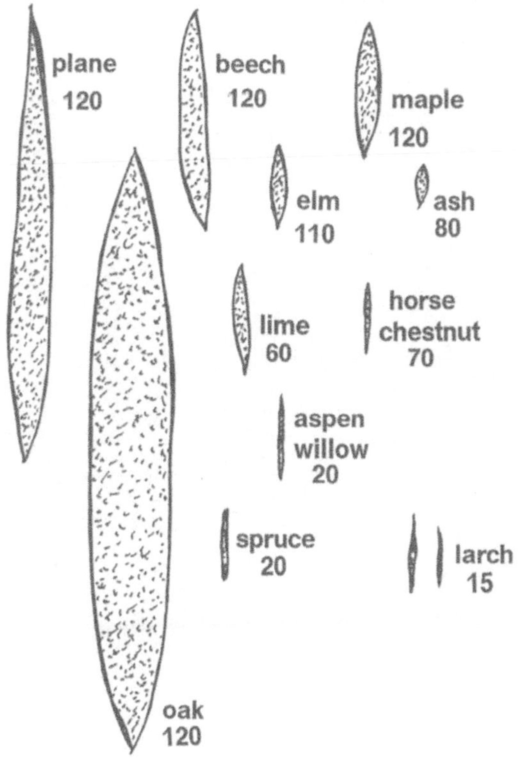

Fig. 95. Some ray cross sections (identical scales) and the corresponding fractometer values

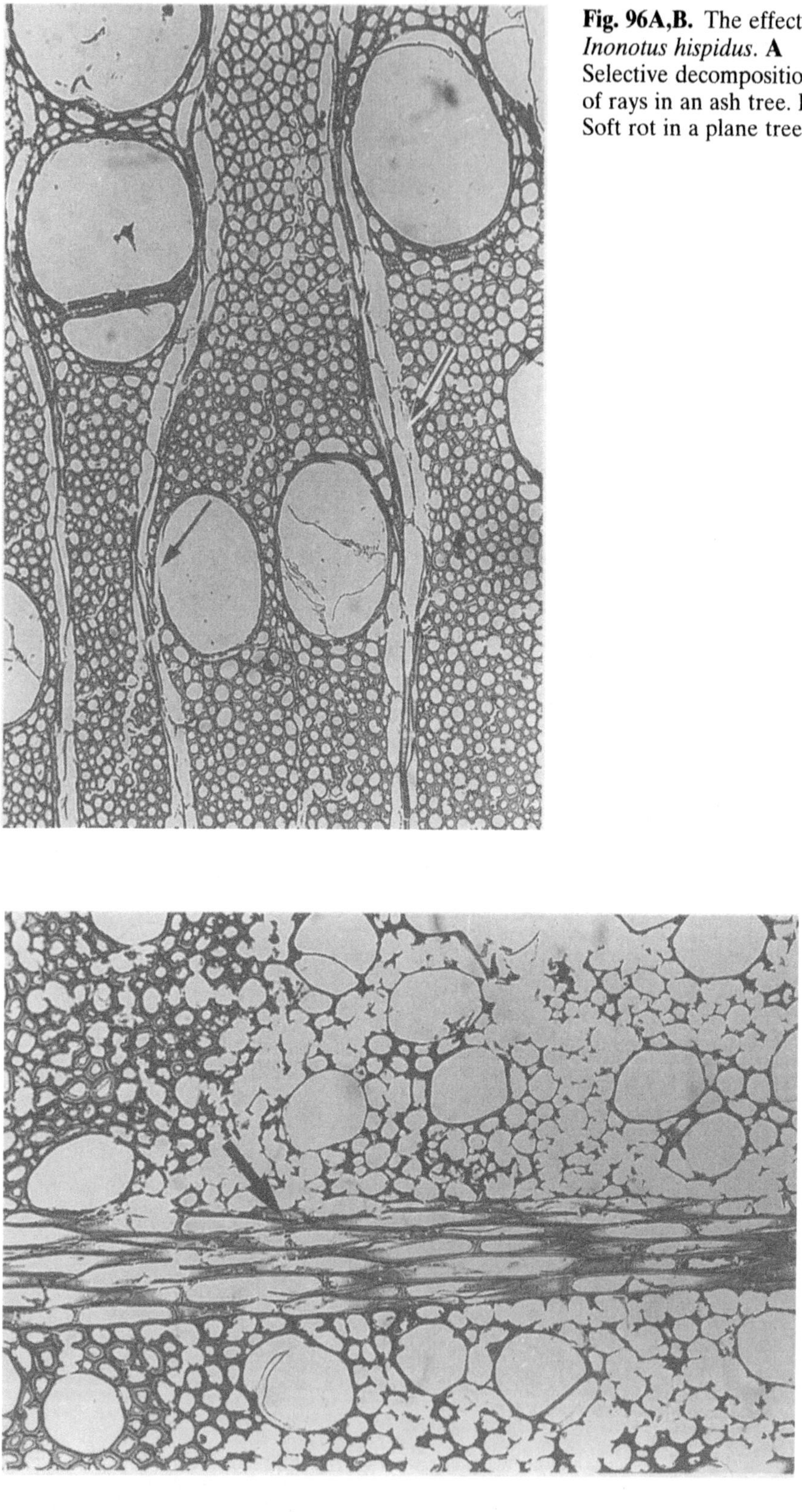

Fig. 96A,B. The effect of *Inonotus hispidus*. **A** Selective decomposition of rays in an ash tree. **B** Soft rot in a plane tree

thicker rays, than for conifers. However, the transverse strengths (or fractometer fracture stresses) of oak trees, as the trees with the largest rays, are not much higher than those of plane trees or beeches. This may be due to the ring porosity of oak and to the additional material that is required in the rays to bridge the "ring cracks" (earlywood).

Extremely low values are found in the literature for radial transverse strengths. Radial transverse strengths are often determined by measuring notched samples, without giving due regard to the notch stresses (Kollmann 1982). Fractometer tests revealed that the radial bending strength of hardwood may reach 50 to 100 % of the values obtained for the axial compressive strength along the grain. This being so, one comes upon a phenomenological explanation of the significance of the rays.

Schwarze (1995) proved the relevance of the rays from a mycological point of view. Fungi of the species of *Inonotus hispidus* selectively decompose the rays of ash trees (Fig. 96), and "disassemble" the ring-porous wood into many small cubes, which collapse easily. The same fungus causes soft rot in

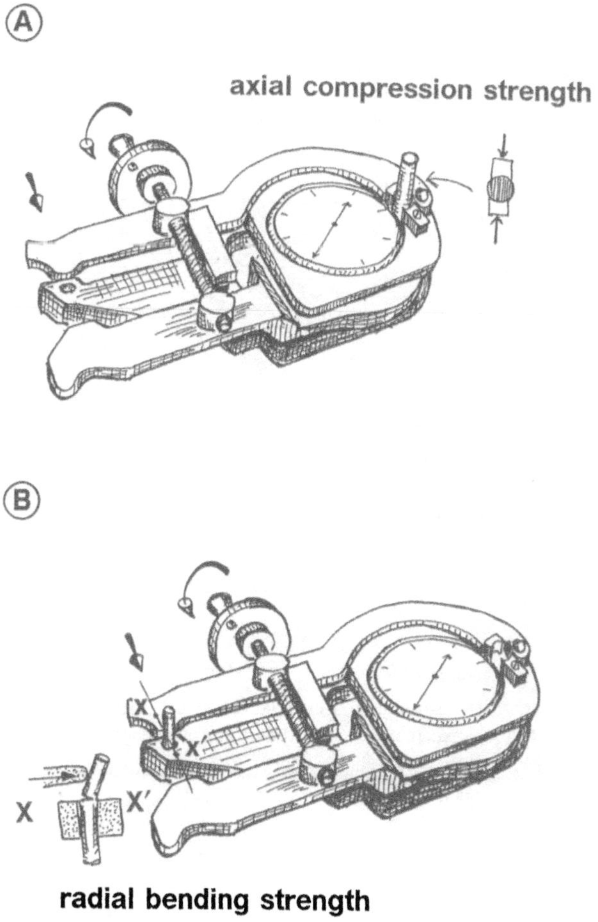

Fig. 97A,B. Extended uses of type II of the fractometer. **A** Compressive-strength measurement along the grain. **B** Radial measurement of the bending strength

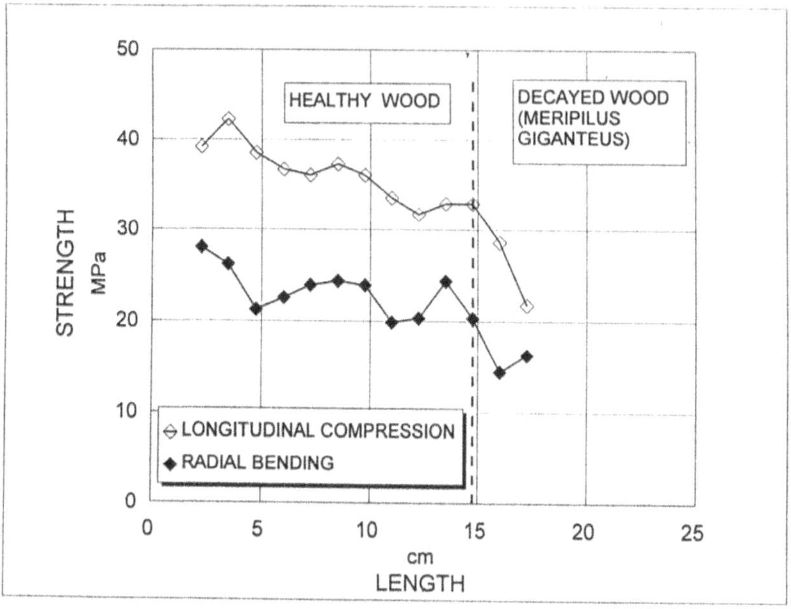

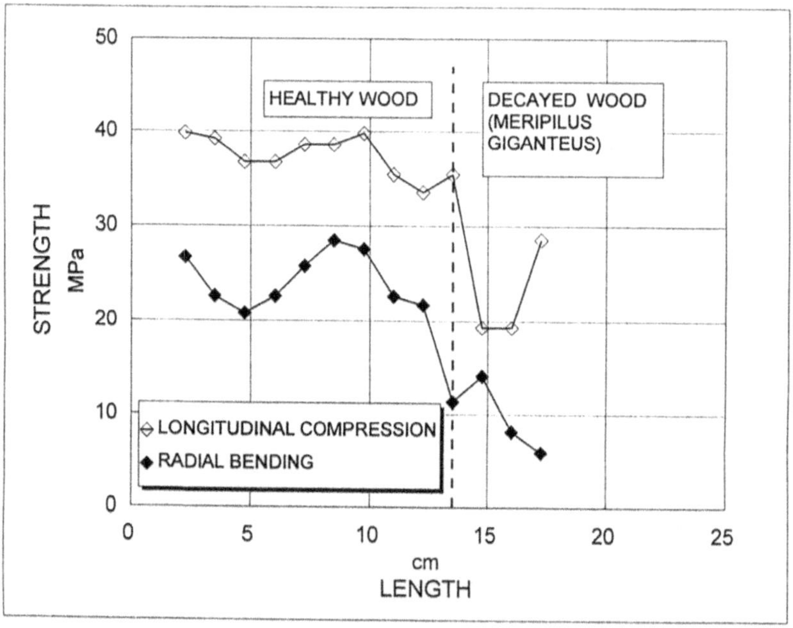

Fig. 98. When trees are infested with fungi, the decrease in their radial bending strength is often greater than the decrease in their longitudinal compressive strength (example: *Meripulus giganteus* in a beech tree)

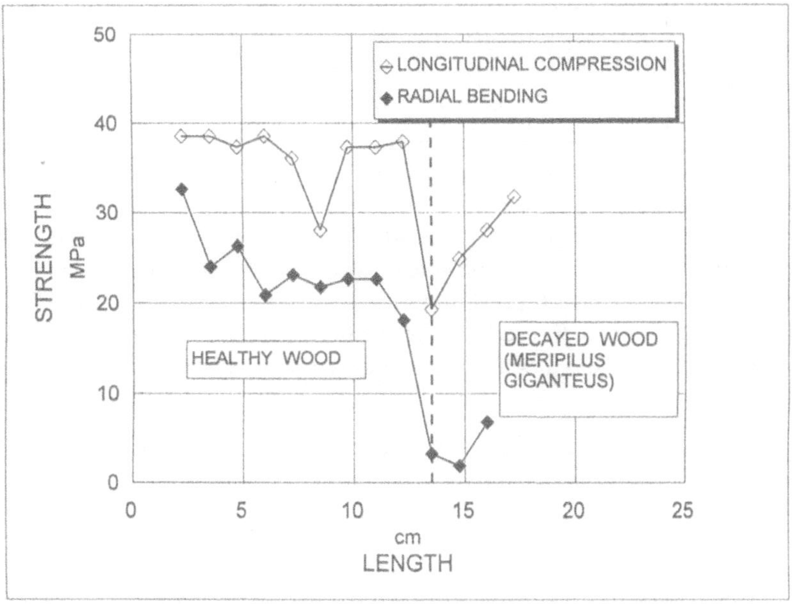

Fig. 98. *Continued*

Platanus acerifolia but is unable to decompose the strongly lignified rays of that plane. *Inonotus hispidus* may infest such plane trees for many years without reducing their transverse strengths. This is also evident regarding the relatively high fractometer values of samples of infested plane and the extremely low values of affected ash wood specimens.

The fractometer has stood the test of evaluating many a suspectedly hazardous tree. However, this device cannot measure axial strengths on upright trees. Fractometer II is an improved type, developed for wood research (Fig. 97). This latest version determines the compressive strength of wood along the grain by squeezing the core transversely (Fig. 97A) by means of special clamps. The values obtained are in surprisingly good agreement with the values compiled by Lavers (1983). Fractometer II measures the radial bending strength of wood (Fig. 97B) by breaking cores in the traditional-fractometer way. The load is indicated by a trailing pointer (see fractometer I). Fractometer II is slightly inferior to fractometer I as regards the lack of angle measurement. Finally, type III is designed to measure all mechanical wood properties shown in Fig. 121.

Fractometer II is used, for example, to assess the degree to which wood-decomposing fungi also reduce the longitudinal strength of wood. Additionally, it is used to determine the correlation between the decrease in longitudinal and lateral strengths. Figure 98 shows the result obtained for *Meripulus giganteus* in a beech tree. In the intact wood, the axial compressive strengths and the radial bending strengths are quite proportional. In decayed wood, both values decrease but the transverse strength often drops more than the longitudinal strength. Evidently, fractometer I provides the more sensitive decay diagnosis.

The fractometer supports wood research by providing a large variety of measured variables and parameters. The internal optimization of trees, i.e., the stress-controlled distribution of transverse strengths, was verified by use of this device. The fact that the highest radial bending strengths are found in the areas of highest transverse stresses (Fig. 99) reveals that wood's resistance to cleavage is greatest in the areas of highest radial cleavage stresses. This was proven through centimeter-by-centimeter breakage of the cores, shown in Fig. 99. The strengths were entered in the FEM model and were plotted for comparison with the FEM radial transverse stresses calculated for bending

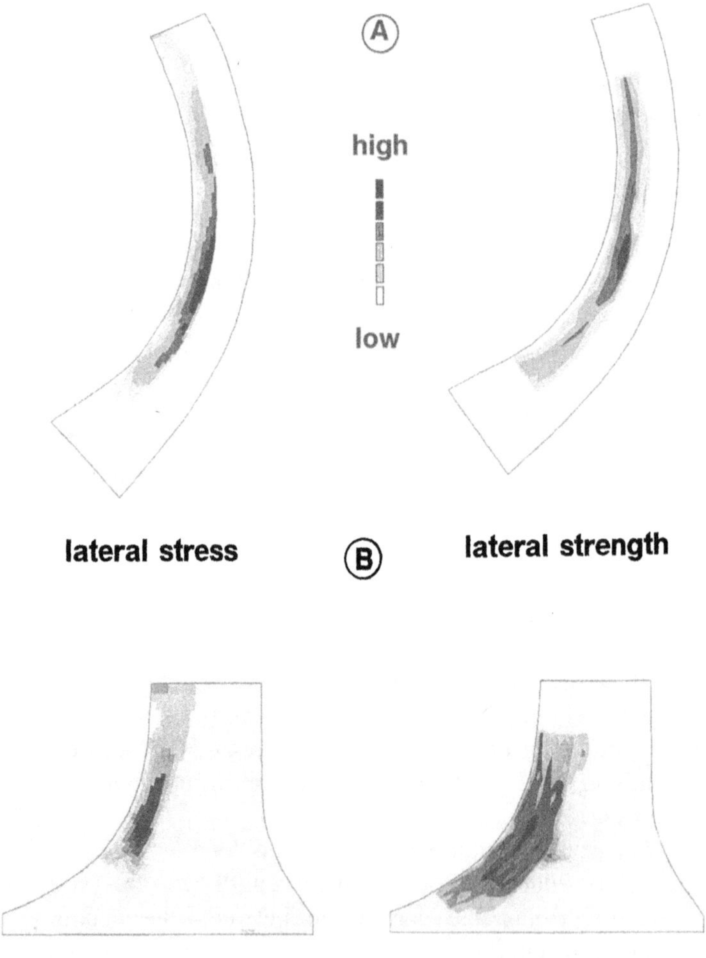

Fig. 99A,B. Load-adapted distribution of radial bending strengths in hazard beams. **A** Saber tree. **B** Root buttress (Results: Wolfgang Albrecht)

loads. The distribution of the radial strengths was found to be largely identical with that of the radial stresses; in other words, the lateral strength of the wood is controlled by lateral stresses. The examples given in Fig. 99A and B represent so-called hazard beams which may cleave lengthwise.

The optimum, load-controlled distribution of tangential strengths is discussed in the section below.

9.3 Tangential Transverse Strengths in Leaning Trees

Tangential bending strengths are determined by examining cores that are removed tangentially (Fig. 100A). It is difficult to measure these values because the fracture surface is parallel to the rays. There is only one point where the lignin matrix is not reinforced by the rays, and it is at this point where tangentially removed cores exhibit their minimum bending strengths when broken (Fig. 100A).

It would certainly be more effective to slice and break radial cores. This radial-core bending method is currently being developed to improve fractometer II, and to determine almost the entire range of relevant measured variables from radial cores (Figs. 100B, 121).

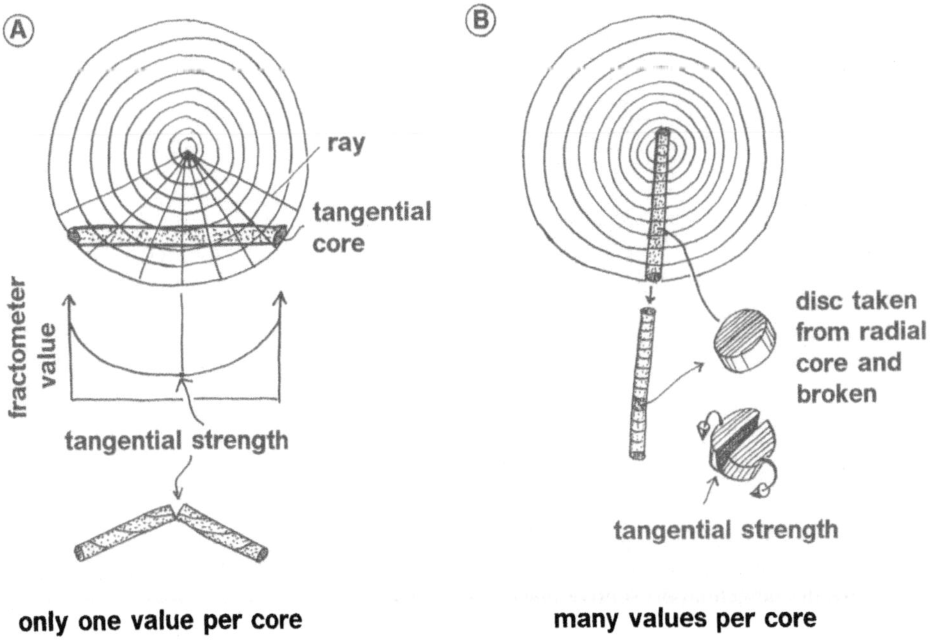

Fig. 100A,B. Measurement of the tangential bending strengths of **A** tangentially removed increment borer cores, **B** radially removed increment borer cores

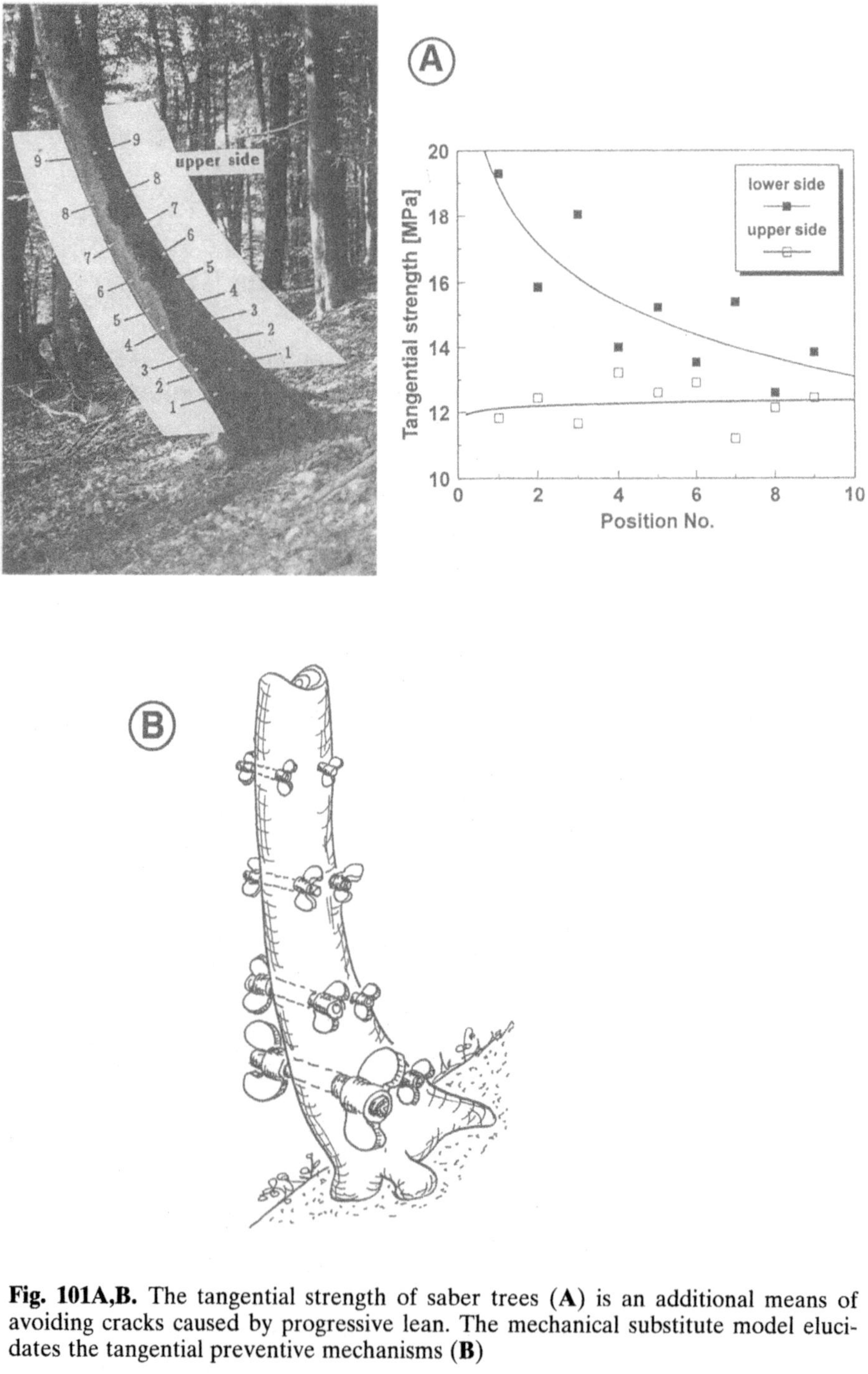

Fig. 101A,B. The tangential strength of saber trees (**A**) is an additional means of avoiding cracks caused by progressive lean. The mechanical substitute model elucidates the tangential preventive mechanisms (**B**)

Saber trees, threatened by cracks caused by progressive lean (Figs. 65, 66), are very well suited for determining optimum strength distributions. A study of both the risk of cracking by the tree and how to prevent it was made by Frank Dietrich and Roland Kappel (Kernforschungszentrum, Karlsruhe) using the fractometer. Different saber trees were tested to prove that the tangential strengths are highest where the bends of the trees are most pronounced (Fig. 101A), in other words where cracks caused by progressive lean are most likely to occur. Recollecting the tangential growth stress properties described above, one must also consider that the bent lower sides of saber trees are compressed by tangential stresses.

Evidently, all areas susceptible to cracking from progressive lean are characterized by both high tangential strengths and compressive growth stresses, which counteract the critical tangential tensile stresses. The model shown in Fig. 101B describes this correlation. The thicker the bolt barrels, the higher the tangential strengths. The larger wing nuts symbolize higher tangential growth stresses. The model shows that the tree counteracts the formation of longitudinal cracks, induced by tangential tension, by locally increasing its tangential tensile strengths. Certainly, this load-adapted improvement in the quality of the wood is also observed in the area of decay-caused cavities.

9.4 Strengths in the Residual Walls of Hollow Trees

The failure of hollow trees is caused by a mechanism that reminds one of that of the failure of a garden hose when it is flattened (Mattheck and Breloer 1995). As the hollow cylinder bends, radial transverse forces develop on the tension and compression sides of bending. These forces ovalize the hollow cross section and cause it to collapse after forming longitudinal cracks (Fig. 102). The fractometer was applied to hollow trees in a field study to investigate whether transverse strengths are higher in the residual wall of the hollow area of the trunk than in the solid trunk above and below. Figure 103 gives a representative example to illustrate the results. The transverse-strength fractometer values obtained for the intact residual walls of hollow trees are higher than those obtained for affected residual walls. Only the weakest trees surrender, using low-grade repair material. Evidently, the fractometer can even evaluate the will of the tree to survive.

Tree failure may depend on the sizes of cavities (Mattheck and Breloer 1993, Mattheck and Bethge, Erb 1993). The maximum permissible hollow spaces were assessed by measuring about 800 hollow trees, either upright or broken. The wall thickness/radius ratio t/R was plotted relative to the stem radius. Figure 104 shows the critical hollow space to be t/R = 0.3. This value was found to apply to the failure of trees with full crowns. Trees whose crowns are reduced may survive with even larger cavities.

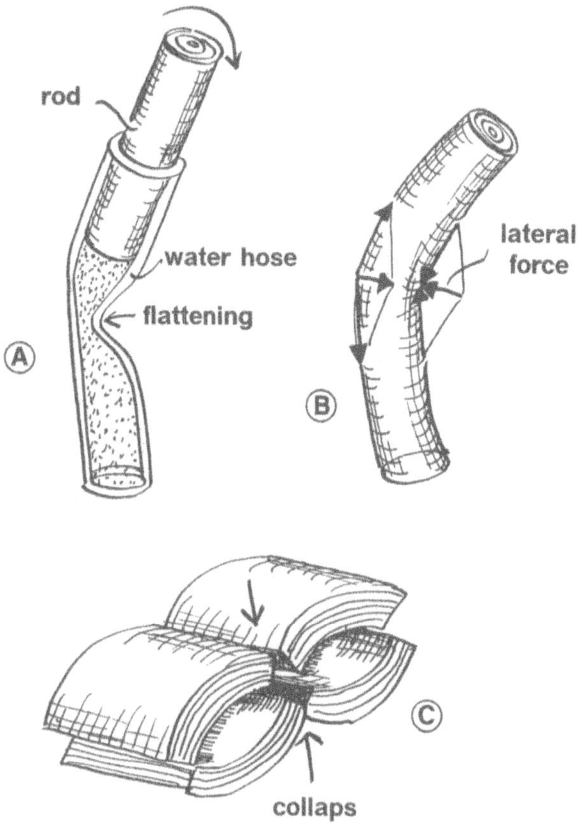

Fig. 102A–C. Cross-sectional flattening may lead to the failure of hollow trees

According to this maximum permissible value, trees whose hollow spaces amount to about 70% of the radius are threatened by failure. This critical value seems not to depend on the species, for it applies equally to strong oak, to weaker poplars or to conifers. The ratio t/R = 0.3 was also found for the tracheids of conifer latewood and for the bones of mammals (Currey and Alexander 1985). Only the walls of bat bones are thinner. Sound orientation may be the compensatory fracture protection for these animals.

Applying Bernoulli's classical bending theory to hollow cylinders, it is easy to see that cavities 0.7·R cause only a moderate increase in stresses upon the surface. The stresses acting on a cylinder whose hollow space amounts to 70% are only about 1.3 times higher than those that act on a solid cylinder. However, cavities >70% cause a rapid increase in stresses. The geometrical critical ratio wall thickness to radius t/R = 0.3, which applies to hollow cylinders subjected to bending, seems to be applicable to the biomechanical principles that rule the flora and fauna generally.

Fig. 103. Increased transverse strengths in the residual wall thicknesses of hollow trees (Results: Wolfgang Albrecht)

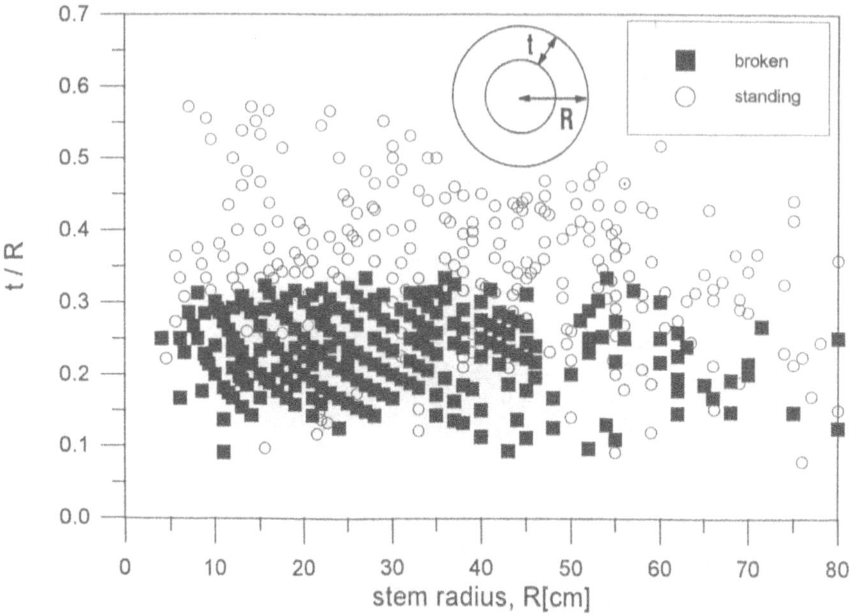

Fig. 104. Critical wall thickness/radius ratios in hollow trees

Summary

- Wood can be imagined as a system of lignin chimneys. Each of the lignin chimneys houses a cellulose vein.
- The rays are the transverse interlocks of this cellulose-reinforced system. They determine essentially the transverse strength of a tree.
- The radial transverse strength is highest in areas of the highest radial tensile stresses.
- The tangential transverse strength is highest in areas of the highest tangential tensile stresses.
- In the case of very vigorous, intact trees, the quality of the repair material is superior to the quality of the wood of parts of the trunk that are not in need of repair.
- Under bending loads, hollow cylinders do not break before their hollow spaces amount to more than 70% of the cylinder radii. This was proved during a field study of different hollow trees.

10 Cracks

10.1 Frost Cracks

The various mechanisms of frost crack formation in trees (Fig. 105) are described by Schirp et al. (1974) and Kubler (1983). In essence, frost cracks develop due to low-temperature shrinking and internal drying.

This chapter discusses the role of the rays as dormant cracks, with regard to low-temperature shrinking. Figures 43 and 44 showed that the spindle-shaped cross sections of the rays account for crack formation during tangential tensile loading. The tangential growth stresses (Fig. 64) compress the rays and reduce the risk of cracking. However, since these tangential stresses are rather low during periods of frost, the tangential tensile stresses caused by low temperatures can scarcely be counteracted, and the rays cleave lengthwise.

Saw cuts exhibiting no tree ring injuries and no enclosed bark or other initial defects indicate that a crack may have been caused by frost alone. Such frost cracks develop ribs at their peripheral ends. Ribs found on the trunk surface are always symptoms of frost cracks, or of other types of cracks just under the surface.

However, the undamaged tree rings typical of frost cracks may also be symptoms of shear-induced cracks or of combined frost and shear cracks.

10.2 Shear Cracks

Shear yielding of the two halves of a trunk (Fig. 106) reminds one of book pages slipping when a book is bent. Yielding is avoided by shear stresses, which are highest in the center of the trunk on the neutral fiber of bending, where bending stresses are zero. Only in this axial plane are the rays subjected to shear instead of to bending. They lose their interlocking properties and behave like roll barrels in a bearing. Evidently, the crack-starting effect of rays is triggered not only by tangential tensile loading but also by longitudinal shear. The rays then turn into slip lines, which run perpendicularly to the bending direction. The shear cracks that develop run perpendicularly, for example south to north under west-to-east wind loads. These shear cracks may either cut entirely through the trunk or may reach only to the trunk center.

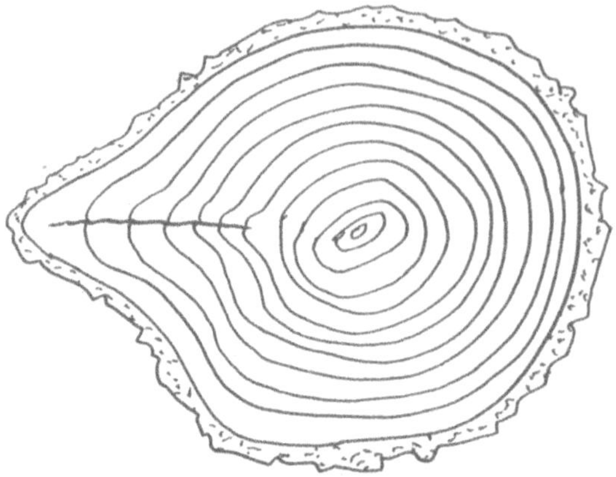

Fig. 105. Frost cracks in a saw cut

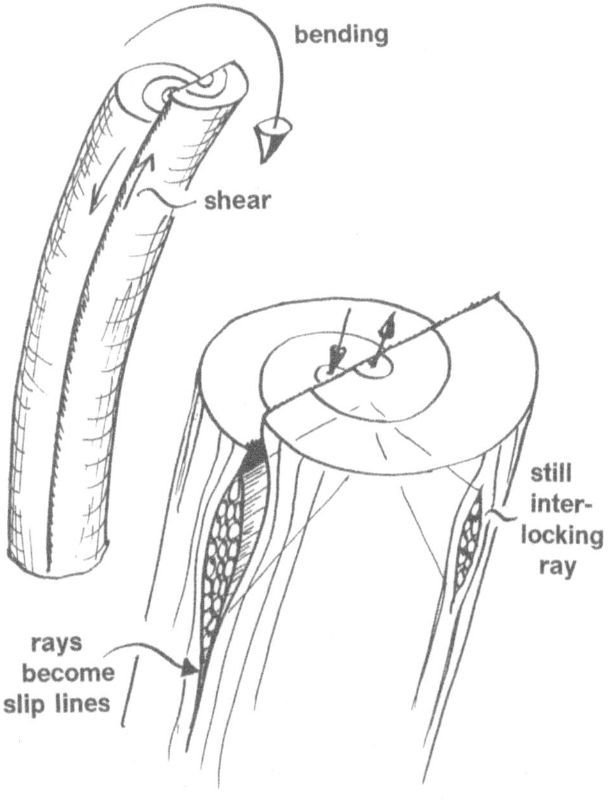

Fig. 106. Shear cracks along the neutral fiber of bending

Bearing this in mind, many supposed frost cracks may actually be shear cracks or combined shear-and-frost cracks. As a rule, one finds only one shear crack running along the ray on the neutral fiber of bending. The interlocking rays which point in the wind direction do not admit any further cracking (Fig. 106).

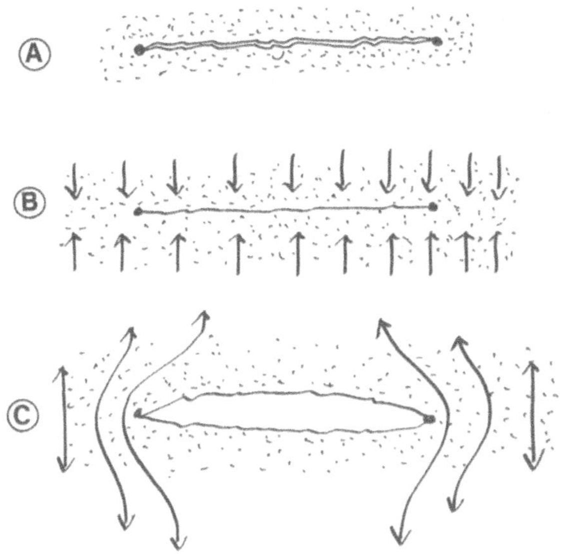

Fig. 107A–C. Cracks are only critical in the case of tensile loading or shear loading

10.3 Enclosed Bark

Compressed cracks are harmless cracks. Figure 107 illustrates how compressive stresses press the edges of cracks against each other so that force-flow diversion a round the cracks is avoided. On the other hand, tension perpendicular to the direction of crack propagation causes the crack to open and effects a sharp force-flow diversion round the crack tips. Cracks are the most hazardous forms of notches, and highest notch stresses are found at the crack tips.

The same applies to enclosed bark. While compressive stresses are transmitted without causing damage, tensile stresses on the contacting bark surfaces convert the enclosed bark into a hazardous crack.

The examples shown in Fig. 108 reveal the large variety of such predetermined breaking points in trees. When exposed to shear stresses or torsional loading, enclosed bark between the root crowns often causes diametrical throughcracks which connect two opposite inclusions. Decapitation of the leading shoot or grafting procedures are followed by growth of new wood over the deadwood, and enclosed bark is found where the repair material is attached. When wires are wrapped by trees, enclosed bark develops where the wood meets and grows over the wire from above and below. Steep branch joints with acute angles between branch and stem are found with inclusions that are not necessarily closed by welds (Chinese beards). In addition to the axial weldings under such Chinese beards, there are cross-weldings with enclosed bark. Cross-weldings with inclusions account for the danger of, for example, girdling roots, which not only strangle the sap flow but also initiate cracking of the highly stressed horizontal root. Wound occlusion, as good and as rapid as it may be, is also always accompanied by enclosed bark (Shigo 1990). When the wound wood coming from both sides is bruised, secondary cracks may start from the edges of the surface of the old wound. Ribs may indicate the presence of enclosed bark even after successful wound occlusion. As a last example, enclosed bark is also found where the stubs of discarded branches are surrounded by trunk wood. Under tensile loading perpendicular to the bark surface, any of these various manifestations of enclosed bark has

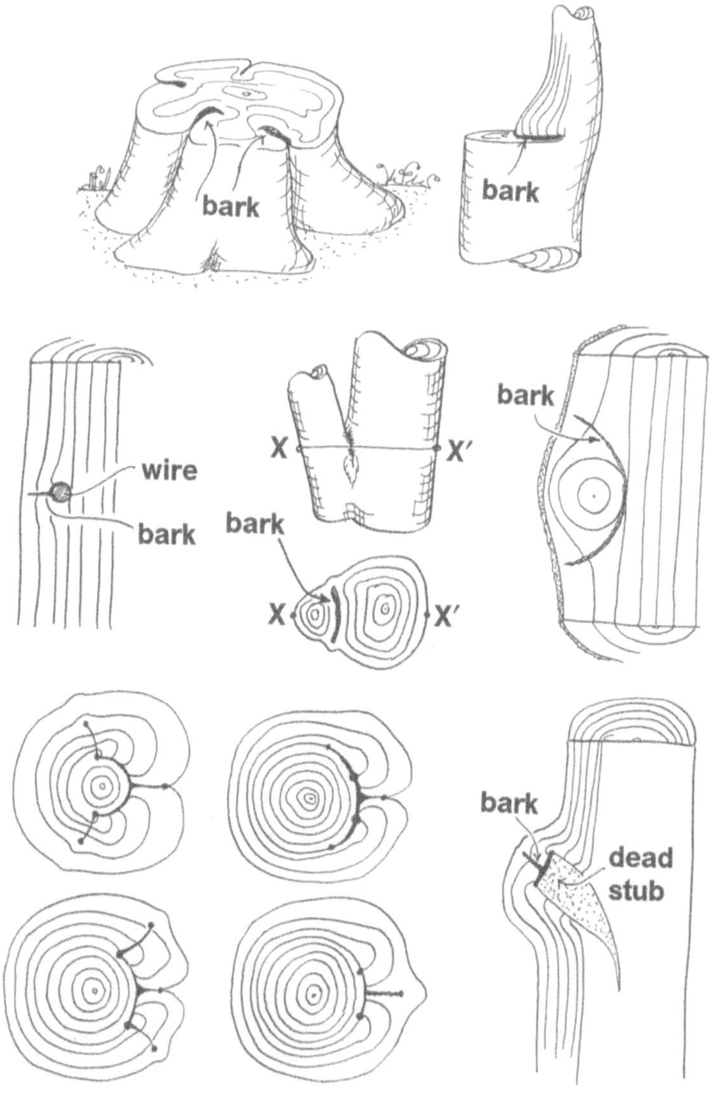

Fig. 108. Different manifestations of enclosed bark

the effect of a crack, and each inclusion may be a critical predetermined breaking point.

10.4 Hazard Beams

Figure 109 illustrates the failure principle of the hazard beam. Hazard beam failure is the reversal of tree failure due to cross-sectional flattening (see Fig. 102). Hazard beams are bent trees or branches that are straightened by bending. The bending stresses, which are diverted due to the presence of the bend, are accompanied by radial tension, which may cause cleavage of the respective tree part. Bent tree parts which are cleaved on the tension side of bending may even split a second or third time until they dangle like ropes from a crane jib. The cracks are arrested by ending in pressure zones on both sides.

The root crowns of spruce are typical hazard beam variants (Mattheck 1991). These root crowns often cleave on the windward side, due to the above-mentioned transverse tensile stresses. Spruce trees mostly fail due to windthrow or through hazard beam cleavage (Fig. 110). The delaminated wood fiber bundle behaves like a loose rope, which straightens under each blast of wind. The trunk swings into this rope at shoulder level and breaks at that level. Evidently, this complicated process is triggered by the root crown hazard beam.

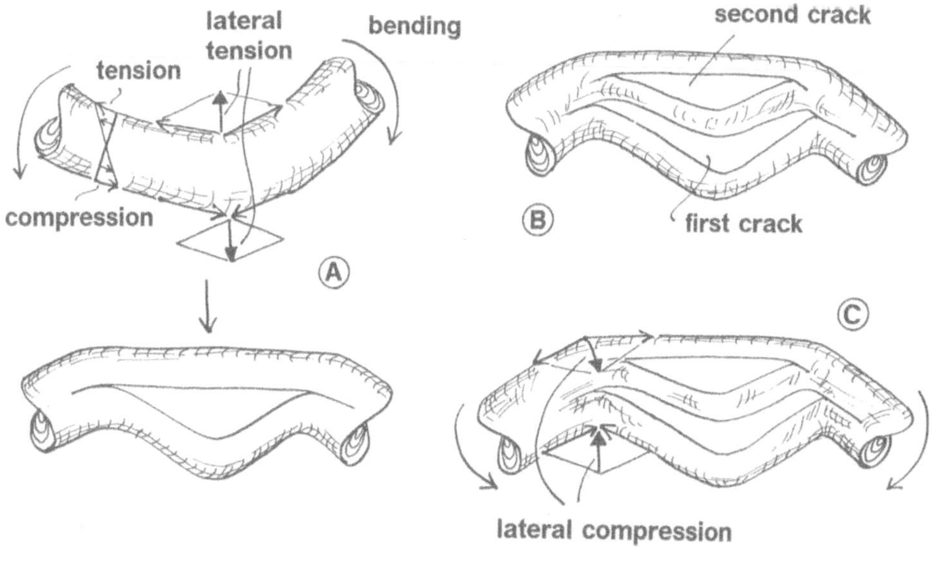

Fig. 109A–C. The hazard beam failure principle. **A** Primary splitting. **B** Secondary splitting. **C** Crack arrest

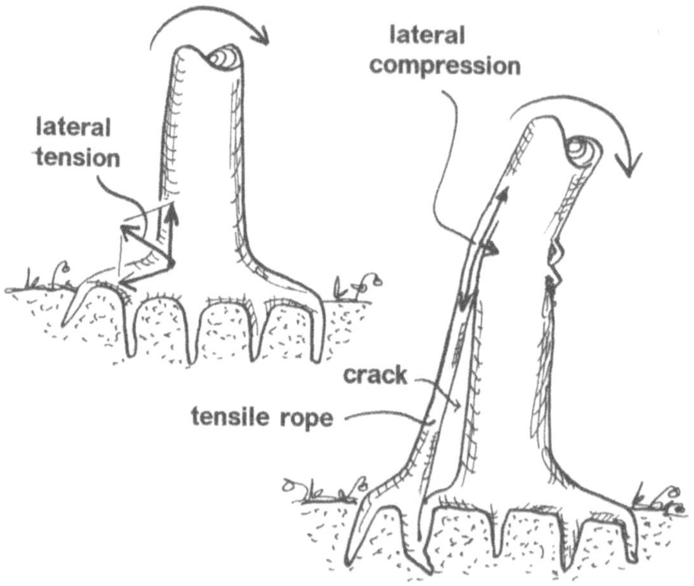

Fig. 110. Hazard beams in the root-to-stem transition areas

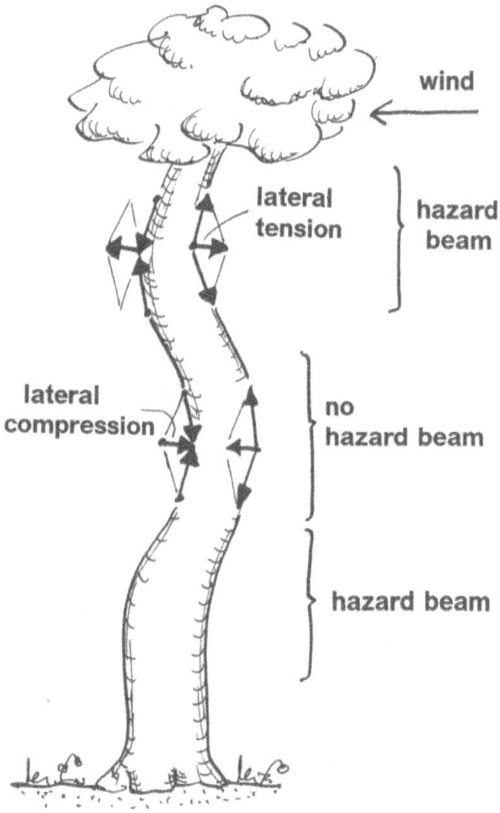

Fig. 111. Hazard beams in meandering trunks

The above does not mean that any bent part of a tree is a hazard beam. Bent trunks and branches become hazard beams only when they are exposed to straightening bending stresses. The meandering trunk shown in Fig. 111 illustrates the typical hazard beam conditions. By no means can the central part of the stem, compressed by transverse forces, be called a hazard beam for this direction of the wind.

Hazard beams could be avoided only by resorting to radical felling and pruning. Figure 112 represents some hazard beams, considering the prevailing wind directions. Under increasing wind loads, these branches would bend up and split lengthwise. A change in the prevailing wind direction would change the hazard beam types and positions. Within trees also, any wound-fiber bundle that is straightened by external forces is a hazard beam in disguise (Fig. 113). Local failure may be induced by, for example, a bend of the local direction of fibers near the rays (Fig. 41) or by the fibril windings around the pits.

Cracks may be caused by

- tension perpendicular to the crack face (mode I),
- shear in the crack plane (mode II),
- antiparallel shear (mode III).

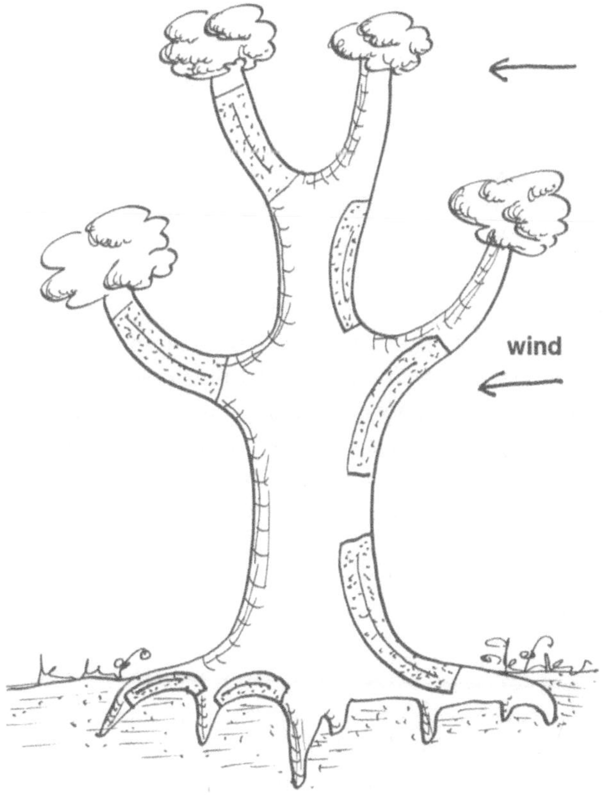

Fig. 112. Hazard beams in branches, and their distribution over the tree

Fig. 113. Disguised hazard beams are found in all sorts of meandering fiber bundles and in any fiber bundle which is straightened by tension

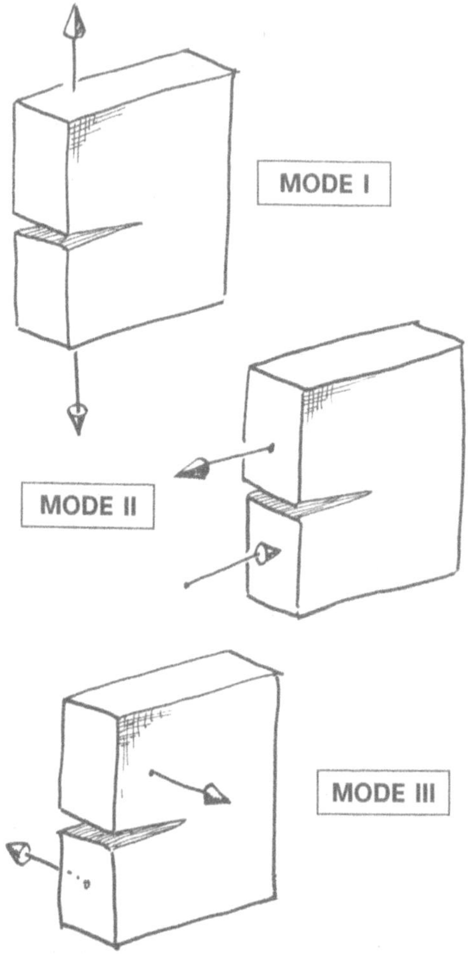

Fig. 114. Classification of cracks in accordance with the principles of fracture mechanics

According to this classification (Fig. 114) hazard beams are mode-I cracks because the lateral tensile stresses act perpendicularly to the crack face. Shear cracks are mode-II cracks. Bark inclusions in acute, narrow tree forks are mode-III cracks whenever the partial stems swing out of the plane of the tree fork while moving towards one another. They are mode-I cracks when the partial stems move away from one another in the fork plane.

10.5 Ring Shakes

The above classification of cracks by modes of loading is particularly relevant to ring cracks or ring shakes, because the modes of loading change as the cracks grow longer. Crack growth may even be stopped due to these varia-

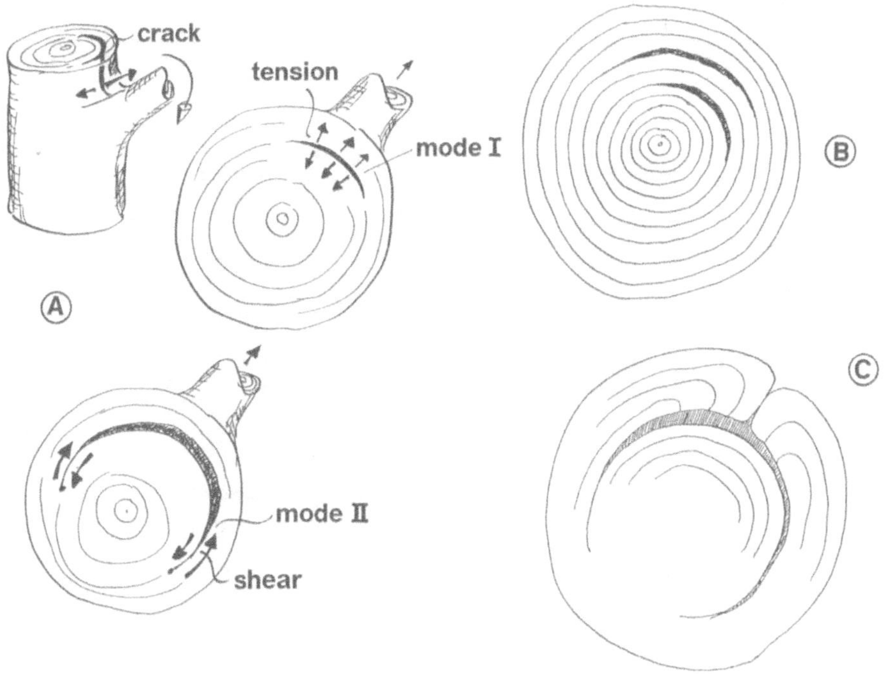

Fig. 115. Ring cracks

tions. The weight of a heavy branch on a ring-porous tree exposes a small section of the large-pore, crack-type earlywood to mode-I loading (Fig. 115A). Mode-II loading is approached, so to speak, as the crack lengthens, with its ends describing a circle defined by the tree ring. The short ring shake is subjected to tension perpendicular to the crack face, whereas the crack ends, which envelop the trunk, are increasingly exposed to shear. If the propagation of the crack around the trunk were not stopped, the crack tips would even be compressed twice. Such cracks are mostly stopped, however, and the branch and surrounding collar are pulled out as the shakes lengthen. Matheny and Clark (1994) list a number of trees whose branches tend to break in the way described above. Most of these trees are of the ring-porous type. They are characterized by single rows of thin rays of low transverse strengths. Ring shakes may also develop due to shear stresses (Fig. 115B) or through radial tensile stresses, as a result of wound wood bruising during wound occlusion (Fig. 115C).

The earlywood of ring-porous trees is a structural weak point, yet the pores of this wood may delay crack propagation. Cracked engineering components that are intended for further operation in spite of their defective condition are pierced right in the middle of the crack tips (Fig. 116A) to enlarge the notch radii. Cracks may be defined as notches with infinitely small notch radii;

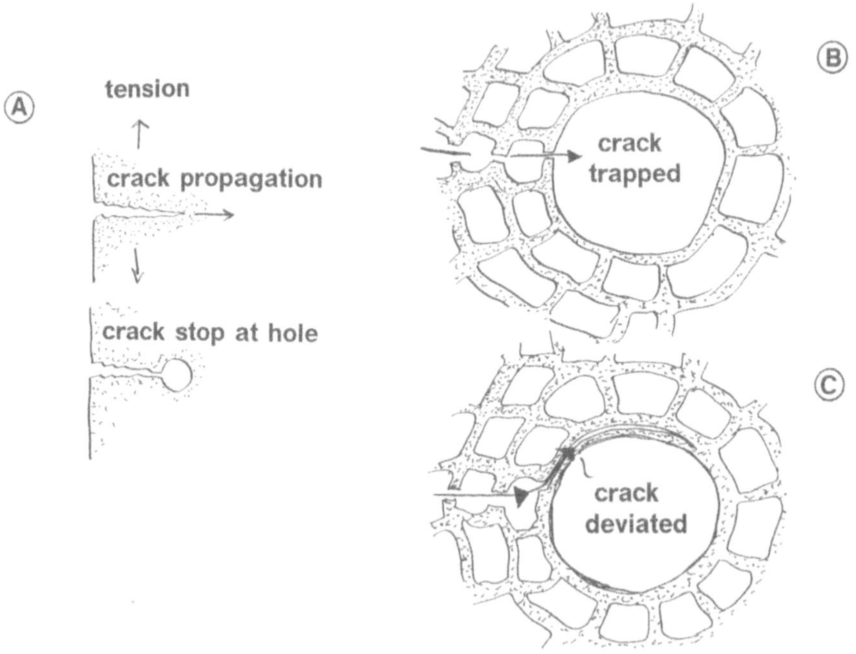

Fig. 116A–C. The vessels as crack stoppers (modified after Gibson and Ashby 1988). **A** Crack growth can stop at holes. **B** Crack capturing in a vessel. **C** A crack is stopped by diversion around a vessel

larger notch radii reduce the notch stresses. Components may be operated for quite a long time before new cracks start from the drilled hole. Sometimes cracks are even stopped permanently.

Cracks that propagate along tree rings may be captured by large vessels. They are "drilled away" in a similar fashion (Fig. 116B). Cracks that run in the walls of vessels, encircling the latter while cleaving the cell wall, depart from worst-case mode-I to approach mode-II loading (Fig. 116C). Gibson and Ashby (1988) published some impressive photographs that prove these two mechanisms.

Earlywood ring porosity may cause crack formation, but can also check crack propagation by capturing (drilling) crack tips, or by diverting crack tips round vessels while approaching less critical mode-II loading.

10.6 Grown Cracks on the Upper Sides of Branch Joints

According to Shigo (1990), branch joints, even those of intact branches, are laminates of stratified stem collars and small barb- or beard-type branch fibers,

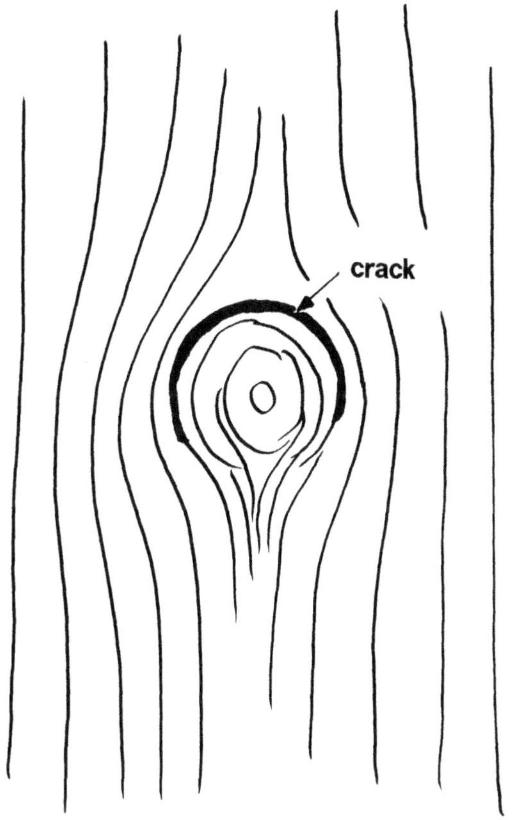

Fig. 117. Natural cracks at the upper side of a branch joint

which are diverted to point downwards (Fig. 117). The fibers carry the force flow and, thus, are diverted following it. The price to be paid for this adapted fiber diversion is the semicircular crack found on the upper side of the branch between branch and stem wood. This crack explains why bent-down branches can be pulled out relatively easily, while bent-up ones resist and remain with the trunk, or if they break, keep their branch joints (Shigo 1990). Evidently, cracks on the upper sides of branches are evolutionary defects that account for the importance of fiber diversion in the direction of the force flow. Nevertheless, branches may be pulled out in the presence of such cracks (Shigo 1990). When the leading shoot is weak or when it is completely missing, the collars round the branch grow weak, and the barb- or beard-type fibers lose their hold as they are diverted.

10.7 T-Cracks

T-cracks are not as exceptional as may be believed, but they are not always large enough to produce identifying ribs in front of the crack tips. T-cracks

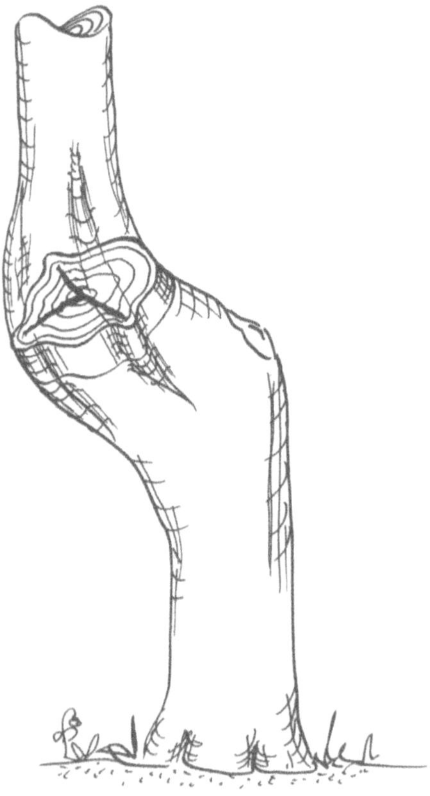

Fig. 118. T-cracks are combinations of "banana" cracks and hazard-beam cracks

develop through combined hazard beam failure and "banana" cracking (Fig. 118). The web of the T represents the crack caused by progressive lean, while the cross-bar symbolizes the hazard beam crack. Certainly, these combined T-shaped cracks occur separately under exactly the same conditions. Both types of cracks occur whenever a bent beam is straightened, and both develop when the rays are insufficiently compressed by tangential growth stresses on the surface of the tree. Under tangential tensile loading, these two conditions suffice to convert the cracks into combined hazard beam/progressive lean T-cracks.

10.8 Drought Cracks

Drought cracks develop due to water losses in excessively hot weather. The low pressure resulting in the tracheids causes the tracheid cross sections to contract, and results in tangential tensile stresses between the walls of adjoining vessels (Fig. 119A). Cracks form at weak points of the cell structure, e.g.,

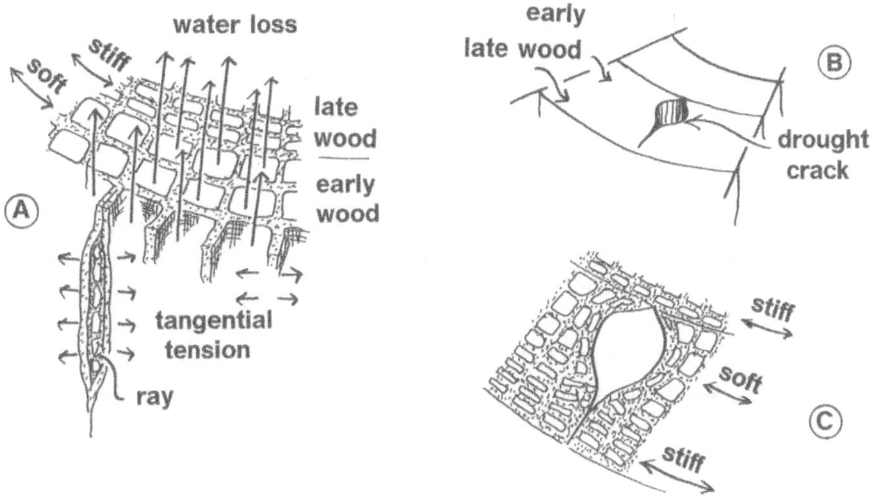

Fig. 119A–C. Drought cracks gape more in soft earlywood

at the rays or at accidental defects in the middle lamella. Unlike the thick-walled, tightly packed latewood, earlywood is soft and elastic when exposed to these tangential tensile stresses. While the thin cell walls and wide cell cross sections of the earlywood admit gaping drought cracks, the stiff latewood has a compressing, retaining effect. Since the thicker-walled cells of the latewood would not contract as much under low pressure as the thin-walled cells, lower tangential tensile stresses can be expected as well. It remains to be investigated whether the resulting rhombus- or diamond-shaped cracks can cause global tree failure. These cracks, however, are microdefects that may accumulate and develop into macro-defects and may impair the service value of wood as a material for supporting structures. In any case, the growth stresses, acting on the tree surface as tangential compressive stresses, counteract the harmful tangential tensile stresses.

11 Final Remarks – Review and Outlook

The failure modes discussed above reveal that even structures as perfectly optimized as wood exhibit vulnerable spots when their optimum state is changed or disturbed. Chapter 3 showed that the external shape and the inner architecture of trees are optimized through the combined action of several self-optimization measures. The failure of one or more of these measures due to external disturbances or decay may inflict the tree with a localized or thorough disaster. The mechanical reserves that the tree can mobilize to avoid such catastrophes are defined and specified by the safety factor.

The optimum state of a tree is determined by five theorems (Fig. 120):

- The principle of minimum lever arms
  The tree minimizes external stresses by reducing the length of the loaded lever arm. Minimum lengths can be achieved through active self-bending of the stiff parts (reaction wood) or through passive yielding of the flexible parts of the tree.
- The axiom of uniform stress
  Unavoidable stresses are distributed evenly on the surface of the tree. Stresses on the surface of the tree are uniform at each point on the time average; grown notches do not cause notch stresses.
- To minimize critical shear stresses, the wood fibers and growth rings are arranged along the axial or tangential force flow.
- The strength of wood depends on the distribution of the mechanical stresses that act in the tree when exposed to external loads. The outer shape of the tree *and* the local internal quality of the wood are optimized and adapted to the degree and type of loading.
- Unavoidable weak points are counterbalanced by the growth stresses developed by the tree. The growth stresses counteract the critical loads that may otherwise cause failure.

These theorems have been explained, and proofs have been furnished in the form of typical examples. Since general evidence and universally valid doctrines cannot be produced, the expert is forced to rely on axiomatic statements. Nature can be neither controlled nor fully understood by man.

Our theoretical argumentation has been founded on a continuum-mechanical basis (finite-element method and experimental measurement), which uses mechanical stresses, Young's moduli, and strains to describe states and processes. The cell wall is the smallest conceivable anisotropic-continuum unit. Certainly, other theoretical approaches would be required for an analysis on the molecular level.

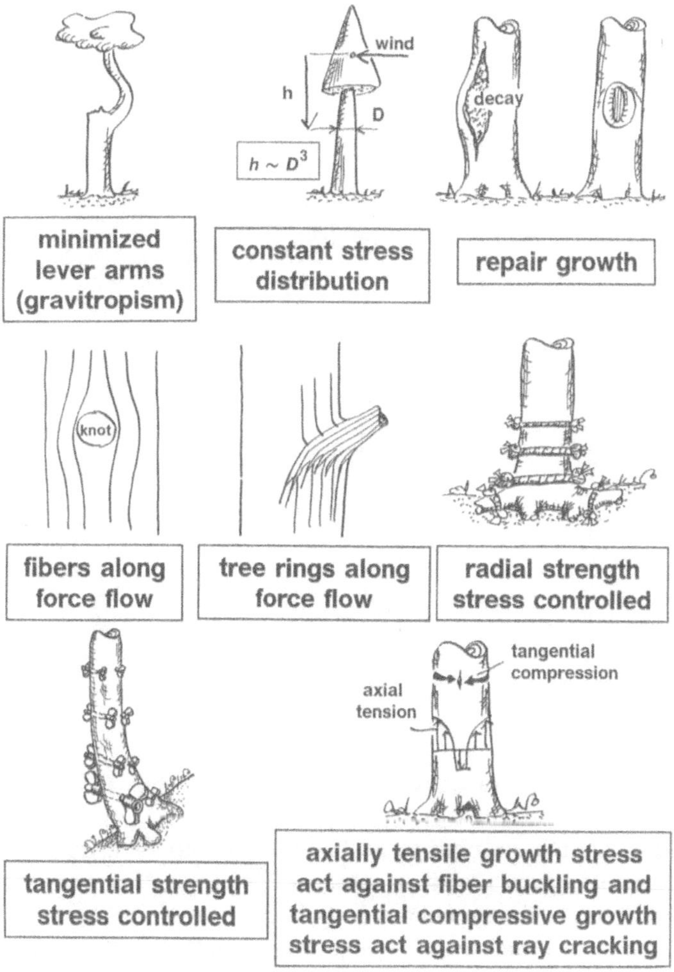

Fig. 120. A survey of the self-optimization of trees

The doubtlessly important cellulose, hemicellulose, and lignin associations (Fengel and Wegener 1984) in the cell walls of wood have not been discussed, although the lignin-hemicellulose linkages may clarify significant properties, such as shock resistance, dynamic shear strength, and cleavage strength. In fact, there might be continuum-mechanical approaches to explaining these adhesive cellulose-lignin mechanisms.

What this book shows is a new approach to determining wood strengths from increment borer cores. The method described can partially replace the intricate, time-consuming test machine procedures applied in the past (Fig. 121). The new method allows one also to determine the strengths of green wood in the open within minutes. Although it cannot measure the

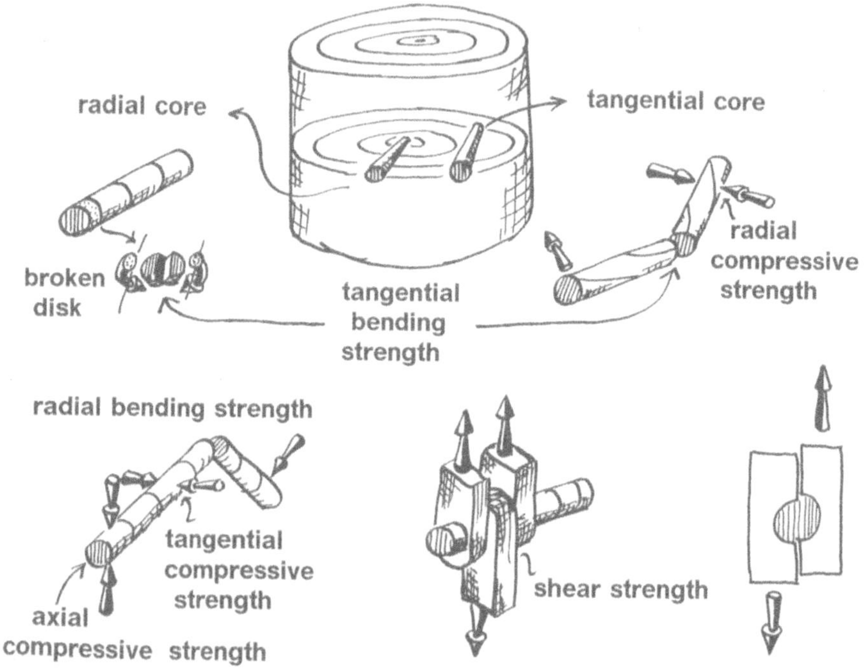

Fig. 121. Strengths determined from increment borer cores by use of fractometer III

effects of large defects (e.g., branches), it is superior to the traditional techniques.

In sum, this book is a first attempt to evaluate the structure of wood in terms of the mechanical optimum that it achieves. The risks that must be expected when parts of this optimum fail are described. The material, wood, is a record of the loads to which the tree was subjected, an authentic biography written in the body language of the tree. This language is understood by anyone able to classify the growth reactions of trees to their relevant mechanical stimuli. Trees are open books, they are diaries of the load histories of ingenious mechanical structures created by nature or possibly . . . of ingeniously machined creatures?

References

Alexander RMcN (1981) Factors of safety in the structure of animals. Sci Prog Oxf 67: 109–130

Archer RR (1987) Growth stresses and strains in trees. Springer, Heidelberg, 236pp

Atherton JC (1978) Experimental techniques for the determination of growth stresses in trees. B S Hon Thesis, Dep For Wildlife Manag Univ of Mass Amherst: 23

Carlquist S (1988) Comparative wood anatomy. Springer, Heidelberg, 436pp

Currey JD (1984) The mechanical adaptations of bone. University Press Princeton, 294pp

Currey JD, Alexander RMcN (1985) The thickness of the walls of tubular bones. J Zool Lond A 206: 453–468

Dietrich F, Mattheck C (1995) Cracks in saber-shaped trees and branches. J Theor Biol 173: 321–327

Fengel D, Wegener G (1984) Wood: Chemistry, Ultrastructure, Reactions. De Gruyter, New York

Gibson LJ, Ashby MF (1988) Cellular solids. Pergamon Press, Oxford, pp 302–303

Harris JM (1989) Spiral grain and wave phenomena in wood formation. Springer, Heidelberg, 214pp

Jacobs MR (1938) The fibre tension of woody stems with a special reference to the genus Eucalyptus. Common For Bur Aust Bull 22, pp 36

Kollmann F (1982) Technologie des Holzes und der Holzwerkstoffe, Springer Verlag, Heidelberg

Kriechbaum R, Schäfer J, Mattheck C (1992) CAIO – computer aided internal optimization. Proc 1st Eur Conf on Smart structures and materials, Glasgow, pp 281–284

Kubler H (1987) Growth stresses in trees and related wood properties. For Abstr 48: 131–189

Kubler H (1991) Function of spiral grain in trees. Trees Struct Funct 5: 125–135

Kubler H (1983) Mechanism of Frost Crack Formation in Trees- A Review and Synthesis. Forest Science 29: 559–568

Lavers G (1983) The strength properties of timber. HMSO, London

Matheny N, Clark J (1994) A photographic guide to the evaluation of hazard trees in urban areas, International Society of Arboriculture, Savoy Illinois USA

Mattheck C (1990) Engineering components grow like trees. Materialwissenschaft und Werkstofftech 21: 143–168

Mattheck C (1991) Trees – the mechanical design. Springer, Heidelberg, 121pp

Mattheck C (1992) Design in der Natur. Rombach, Freiburg, 242pp

Mattheck C (1993) Design in der Natur. Rombach, Freiburg, 242pp

Mattheck C (1995) Wood- The internal Optimization of Trees. Arboricultural Journal 19: 97–100

Mattheck C, Bethge K (1993) Das Fractometer. Allg. Forstztg. 48: 115–117

Mattheck C, Bethge K, Erb D (1993) Failure criteria for Trees. Arboricultural Journal 17: 201–209

Mattheck C, Bethge K, Schäfer J (1993) Safety factors in trees. J Theor Biol 165: 185–189

Mattheck C, Breloer H (1993) Handbuch der Schadenskunde von Bäumen. Rombach Verlag, Freiburg

Mattheck C, Breloer H (1995) The body language of trees. HMSO, London

Mattheck C, Schwarze F (1994) Die Holzstrahlen als getarnte I- Balken in einem mechanischen Ersatzmodell für Holz. Allg. Forst- u. Jagdztg. 10/11: 197–201
Mattheck C, Walther F (1991) Eine neue Fälltechnik zur Vermeidung von Hirnrissen in Laubbäumen. Allgemeine Forst und Jagdzeitung 162: 181–185
Metzger K (1893) Der Wind als maβgeblicher Faktor für das Wachstum der Bäume. Münd Forstl Hefte Springer (Berl) 3: 35–86
Mosbrugger V (1990) The tree habit in land plants. Springer, Heidelberg, p 46
Neef F (1914) Über Zellumlagerung. Ein Beitrag zur experimentellen Anatomie. Z Bot 6: 465–547
Nicholson JE (1971) A rapid method for estimating longitudinal growth stresses in logs. Wood Sci Technol 5: 40–48
Peterson RE (1974) Stress concentration factors. Wiley, New York, 317pp
Polge H, Thiercelin F (1979) Growth stress appraisal through increment core measurement. Wood Sci 12: 86–92
Schirp M, Kubler H, Liese W (1974) Investigations on tree rounds about formation of frost cracks. Forstwiss Centralbl 93: 127–136 (in German)
Schwarze F (1995) Entwicklung und biomechanische Auswirkungen von holzzersetzenden Pilzen in lebenden Bäumen und in vitro. SVK-Verlag, Erndtebrük
Schwarze F, Mattheck C, Breloer H (1993) Der spröde Baumbruch-verursacht durch den Brandkrustenpilz, Neue Landschaft 38: 737–747
Schweingruber FH (1990) Anatomie europäischer Hölzer. Haupt, Bern, 800pp
Shigo AL (1990) A new tree biology. Shigo & Trees, Durham, 618pp
Shigo AL (1994) Tree anatomy. Shigo & Trees, Durham, 104pp
Tantichaiboriboon V, Cook RD (1976) Effect of shape of cut on growth stress induced cracking in cut timber. Rep CEEM-76-101 Coll Eng Univ Wisc Madison, 245pp
Timell TE (1986) Compression wood in gymnosperms. Springer, Heidelberg, 2150pp
Troll W (1959) Allgemeine Botanik, Enke Verlag, Stuttgart
Wagenführ R (1989) Anatomie des Holzes, 4te Anfl. VEB Fachbuchverlag, Leipzig, p 174

Subject Index

GPSR Compliance

The European Union's (EU) General Product Safety Regulation (GPSR) is a set of rules that requires consumer products to be safe and our obligations to ensure this.

If you have any concerns about our products, you can contact us on ProductSafety@springernature.com

In case Publisher is established outside the EU, the EU authorized representative is:

Springer Nature Customer Service Center GmbH
Europaplatz 3
69115 Heidelberg, Germany

Batch number: 10761599

Printed by Printforce, the Netherlands